Universität
Augsburg
University

TECHNISCHE
UNIVERSITÄT
MÜNCHEN

THE GEORGE
WASHINGTON
UNIVERSITY

WASHINGTON, DC

MAX-PLANCK-GESELLSCHAFT

MIPLC Studies
Edited by

Prof. Dr. Christoph Ann, LL.M. (Duke Univ.)
TUM School of Management

Prof. Robert Brauneis
The George Washington University Law School

Prof. Dr. Josef Drexl, LL.M. (Berkeley)
Max Planck Institute for Innovation and Competition

Prof. Dr. Michael Kort
University of Augsburg

Prof. Dr. Thomas M.J. Möllers
University of Augsburg

Prof. Dr. Dres. h.c. Joseph Straus
Max Planck Institute for Innovation and Competition

Volume 32

Qinghua Yang, Ph.D.

Aegis or Achilles Heel:
The Dilemma of Homology in
Biopatents in the Wake of Novozymes

Nomos

MIPLC Munich Augsburg
Intellectual München
Property Washington DC
Law Center

The Deutsche Nationalbibliothek lists this publication in the
Deutsche Nationalbibliografie; detailed bibliographic data
are available on the Internet at http://dnb.d-nb.de

a.t.: Munich, Master Thesis Munich Intellectual Property Law Center, 2017

ISBN   978-3-8487-5021-4 (Print)
         978-3-8452-9271-7 (ePDF)

British Library Cataloguing-in-Publication Data
A catalogue record for this book is available from the British Library.

ISBN   978-3-8487-5021-4 (Print)
         978-3-8452-9271-7 (ePDF)

Library of Congress Cataloging-in-Publication Data
Yang, Qinghua
Aegis or Achilles Heel: The Dilemma of Homology in Biopatents in the Wake
of Novozymes
Qinghua Yang
72 p.
Includes bibliographic references.

ISBN   978-3-8487-5021-4 (Print)
         978-3-8452-9271-7 (ePDF)

1st Edition 2018
© Nomos Verlagsgesellschaft, Baden-Baden, Germany 2018. Printed and bound in Germany.

Acknowledgements

In the past a couple of months, I have been working on the *support* requirement for homology claims. In this thesis, it is *support* that salvaged Novozymes' patent. Outside the thesis, it is also *support* that accompanied me throughout this rewarding journey at MIPLC.

The *support* came from my thesis advisor Prof. Joseph Straus. It is his wisdom and insights that enlightened my interest in the patent law issues in biotechnology. It is his carefulness and patience that guided me through the mist of thesis writing.

The *support* came from Mrinalini and Seth who did their best to take good care of the students, coordinate this programme, and facilitate the teaching and learning.

The *support* came from Yuan who oriented me in Munich, provided me the first-hand experience, and tolerated most of my nonsense. The *support* came from Lan who shared this journey and worked together with me, nine to nine, in the library.

The *support* came from Nadiya who helped me in finalising the last publication of my earlier research. The *support* came from my besties Bahar and Carolina who shared the hobby with me and dispersed my loneliness. The *support* also came from the other colleagues of Class 2016/17 whom I could not enumerate here.

The *support* came from Takeshi and Jingdong who furnished me valuable knowledge and skills from their patent examination practices.

Lastly and most importantly, the *support* came from my family who were always at the back of me, and gave me the faith to explore my life.

At the final moment of this LL.M. programme, I would like to express my sincere gratitude to my thesis advisor, friends and colleagues. You made my journey at MIPLC fruitful. And you made the programme at MIPLC memorable. Thank you.

Table of Contents

Abstract

Biological inventions frequently involve polypeptides, proteins and nucleic acids. Sequences of these molecules are disclosed for patent application. To obtain a broader scope of protection, an applicant employs homology language to formulate the claims and create a homology range surrounding the disclosed sequence. This homology range encompasses sequences that are expected to perform similar functions as the disclosed one does. However, the homology claims face a hurdle that they may not be supported by the written description. In a recent case, *Novozymes*, the Supreme Court of China ruled that homology claims lack support, but a further limitation by species of origin could satisfy this requirement. In this thesis, it is found that species of origin is not an effective limitation. Homology, as the essence of the dispute in *Novozymes*, should have been adequately addressed by the courts. Homology dictates the skilled person's confidence on the functionality of unknown sequences, and is involved in multiple patentability requirements. Therefore, the assessment of support concerning homology shall not be isolated from other patentability requirements. An empirical study shows that the current views on homology are different in the requirements of inventive step and support, thus creating an unclaimable gap along homology values. This gap may constitute a discrimination to biotechnology. This thesis shows that the disparity in views on homology is caused by intermingling the requirements of sufficient disclosure and support. To fix this problem, an appropriate test is furnished for assessing the support requirement concerning homology claims. It may help to narrow the unclaimable gap, meanwhile avoiding prejudice to other inventions. A more reasonable scope of protection is expected to be conferred to sequence-related biological inventions in the future.

Acronyms and Abbreviations

AA	Amino Acid
the Court	the Supreme People's Court of the People's Republic of China
DNA	Deoxyribonucleic Acid
EPC	European Patent Convention
EPO	European Patent Office
EWHC	the High Court of Justice of England and Wales
HFCS	High Fructose Corn Syrup
HL	House of Lords
JPO	Japan Patent Office
Paris Convention	Paris Convention for the Protection of Industrial Property
the Patent Law	Patent Law of the People's Republic of China
PRB	Patent Reexamination Board
RNA	Ribonucleic Acid
SIPO	State Intellectual Property Office of the People's Republic of China
TBA	Technical Board of Appeal
TRIPS Agreement	Agreement on Trade-Related Aspects of Intellectual Property Rights
UKIPO	Intellectual Property Office of the United Kingdom

Chinese Document Nomenclature

Note: Chinese document identifiers are searchable as cited.

A brief translation of the Romanised Chinese characters is provided below:

Er Zhong Min San Chu Zi	First Instance Case, Civil Ligation, by the Third Chamber of the [place] Second Intermediate People's Court
Fa Shi	Judicial Interpretation Document issued by the Supreme People's Court
Gao Xing (Zhi) Zhong Zi	Final Instance Case, Administrative Litigation on Intellectual Property Law Matters, by the [place] High People's Court

Guo Fa	Official Document issued by the State Council
Jin Gao Min San Zhong Zi	Final Instance Case, Civil Litigation, by the Third Chamber of the Tianjin High People's Court
Yi Zhong Zhi Xing Chu Zi	First Instance Case, Administrative Litigation on Intellectual Property Law Matters, by the [place] First Intermediate People's Court
Zui Gao Fa Xing Zai	Retrial Case of Administrative Litigation by the Supreme People's Court

I. Introduction

A biological sequence is a single and continuous molecule, either nucleic acid or protein, presented in the form of its structural combination. Sequences are the essential substance of many biological inventions. The functionality of such inventions is predominantly determined by the order of 4 nucleobases- designated as C, G, A, and T (U), in the case of DNA (RNA), or 20 amino acid residues- designated as single or triple alphabet(s) codes in the event of proteins, such as R or Arg for arginine. Changes to these sequences may lead to disparate results: from complete loss-of-function to approximately maintaining identical functions. The phenomena of codon degeneracy (for nucleic acids) and neutral mutation (for proteins) constitute the main basis of shared functions amongst similar biological sequences, or homologous sequences.

With the state-of-the-art biotechnology, mutagenesis[1] proves easier each day at an unprecedented pace. Modification of biological sequences and making variants become relatively simple tasks. The same functionality of a certain biological sequence is assumed to be easily achieved by creating a variant imitating the reference sequence. Moreover, the possible number of variants can easily reach an astronomical figure according to combinatorics. As a consequence, patent protection over only the specific sequences disclosed in a patent could not reward the contribution of its inventor, and thus cannot achieve the *quid pro quo* of the patent system. Inventors, therefore, wish to draft their patent claims in a broader way so as to encompass a wide range of similar sequences, usually by means of a minimal percentage homology to a specific sequence. On the other hand, this practice may bring anti-innovation effects, as some technical progress owing nothing to the teachings of these patents could also fall within the claimed scope of protection, merely because of sequence homology. Therefore, a delicate line should be drawn as to what extent the inventors are allowed to claim homologous sequences.

---

1 See *Mutagenesis* at Wikipedia <https://en.wikipedia.org/wiki/Mutagenesis> accessed 10 September 2017. "Mutagenesis is a process by which the genetic information of an organism is changed, resulting in a mutation".

Homology claims are generally allowed by patent offices across many jurisdictions, exemplified in examination guidelines by UKIPO.[2] However, the allowable threshold may vary from case to case, and may also change from time to time.[3] It reflects the plight of patent offices in balancing the interest of patent proprietors and the public. A recent case in China, *Novozymes*,[4] has brought homology claims in hot water again. Briefly, a patent relating to one kind of thermostable glucoamylase was invalidated by the Beijing High Court for lack of support,[5] due to the homology language used in the claims. It was argued that the homology claims encompass a large number of variant sequences whose functionality cannot be predicted, and a person skilled in the art cannot reasonably know which particular variant would work the invention. The Supreme Court upheld the patent on the ground that a further "species of origin" limitation in its auxiliary claims narrows down the scope of protection to only a few sequences attributed to a particular species, and a skilled addressee would reasonably predict that the sequences within the same species perform functions similar to each other.

Although the Supreme Court's decision stabilised the patent-in-suit on a seemingly valid ground, it did not touch upon the essence of the issue, and could possibly leave in problems for the future. This thesis aims at discussing the role of homology in biotech patents particularly in relation to

---

2 Intellectual Property Office of the UK, *Examination Guidelines for Patent Applications relating to Biotechnological Inventions in the Intellectual Property Office* (6 May 2010, last updated: 21 October 2016) (UK Biotech Guidelines) 49, Example 5: "A protein / polypeptide having the sequence SEQ ID No. 1 or a variant, homologue, or portion / fragment thereof".

3 Ibid. "There is no general rule for determination of the required agreement, which depends on context, most significantly the stringency conditions. As an example, a low homology sequence may 'pick out' a newly sequenced DNA/RNA, whereas to separate sequences encoding isoenzymes (which have closely related structures), homology of over 95% may be required. Thus the scope of the claim needs to be considered in the context of the specification as a whole".

4 *The Patent Reexamination Board (PRB) & Novozymes A/S v. Jiangsu Boli Bioproducts (Boli) Co., Ltd*, the Supreme People's Court (2016) Zui Gao Fa Xing Zai No.85; *The Patent Reexamination Board & Novozymes A/S v. Shandong Longda Bioproducts Co., Ltd (Longda)*, the Supreme People's Court (2016) Zui Gao Fa Xing Zai No.86.

5 *Novozymes v PRB*, The Beijing High People's Court (2014) Gao Xing (Zhi) Zhong Zi No. 3522; *PRB v Boli, Longda* The Beijing High People's Court (2014) Gao Xing (Zhi) Zhong Zi No. 3523/3524.

polypeptides and proteins, analysing the drawbacks of the *Novozymes* decision, and providing suggestions for future patent practice.

In this thesis, Section II summarises the ins and outs in relation to *Novozymes*. In the same part, reasons for the decision not being satisfactory are presented. Section III discusses the meaning of homology language in relation to its technical background. Next, Section IV explains that *Novozymes* leaves an unjustifiable and unclaimable gap in the technological space under patent law. Finally, Section V presents a more appropriate way to apply the test for support requirement.

II. Novozymes – a Long and Hard Journey to Patent Validity

The *Novozymes* decision was delivered by the Supreme Court of China on 31 Dec 2016. It dealt with the patent validity dispute dating back to 2011, when a parallel patent infringement case was on trial.[6] This section will describe the patent and related proceedings in detail, followed by the author's analysis of the merits of this final decision.

A. The Glucoamylase

Glucoamylase is one of the widely used bio-catalysts in the food industry. It has traditionally been produced by employing filamentous fungi, like *Aspergillus niger* and *Aspergillus awamori*. Glucoamylase is an exo-acting amylase catalysing the release of D-glucose from the non-reducing ends of starch and related oligo- or polysaccharide molecules (see Figure 1).[7] D-glucose is an essential substrate for a number of fermentation processes and for a range of food and beverage industries.[8]

---

6 *Novyzymes v Longda*, The Tianjin Second Intermediate People's Court (2011) Er Zhong Min San Chu Zi No. 81; *Novozymes v Boli*, The Tianjin Second Intermediate People's Court (2011) E Zhong Min San Chu Zi No. 82; *Longda v Novozymes*, The Tianjin High People's Court (2012) Jin Gao Min San Zhong Zi No.41; *Boli v Novozymes*, The Tianjin High People's Court (2012) Jin Gao Min San Zhong Zi No.42.

7 See Julia Marín-Navarro and Julio Polaina, 'Glucoamylases: Structural and Biotechnological Aspects' (2011) 89 Applied Microbiology and Biotechnology 1267.

8 Pardeep Kumar and T Satyanarayana, 'Microbial Glucoamylases: Characteristics and Applications' (2009) 29 Critical Reviews in Biotechnology 225 <http://www.tandfonline.com/doi/full/10.1080/07388550903136076> accessed 10 September 2017.

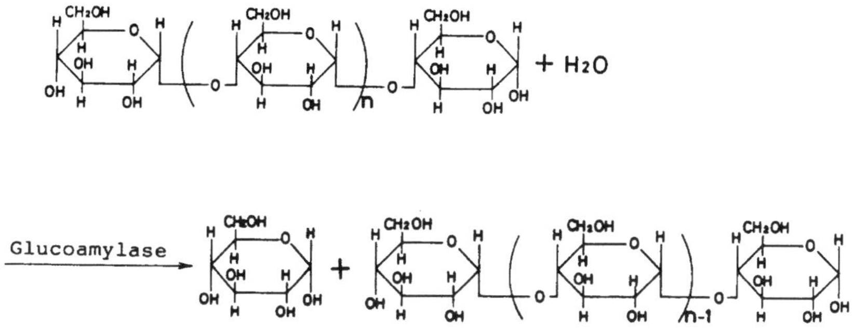

Figure 1. Glucoamylase-catalysed hydrolysis of terminal (1->4)-linked alpha-D-glucose residues successively from non-reducing ends of the chains with release of beta-D- glucose.[9]

An important application of glucoamylase is in the production of the commonly-used high fructose corn syrup (HFCS). Glucoamylase is employed to convert partially-hydrolysed corn starch by α-amylase to glucose, which is further converted by glucose isomerase to a mixture composed of glucose and fructose. This type of mixture, often further enriched with fructose, is commercialised as HFCS in worldwide trades. The HFCS is the largest tonnage product produced by an enzymatic process, making glucoamylase one of the most important industrial enzymes only second to protease. [10]

Ideally, it is economically advantageous if the three enzymes in the catalytic process share the same working conditions. In such way, the enzymatic reactions can proceed without changing vessels and ambient parameters to adapt each enzymatic process. However, the *Aspergillus* glucoamylase has certain limitations, such as moderate thermostability and acidic pH conditions, which increase the cost of the catalytic process. Accordingly, the search for new glucoamylases of optimal pH and temperature have been major goals of research over the years.

---

9 Source: A Kariyone, Y Hashizume and R Hayashi, 'Enzyme Electrode for Measuring Malto-Oligosaccharide and Measuring Apparatus Using the Same' <http://www.google.com/patents/EP0335167A1?cl=en> accessed 10 September 2017.

10 Vimal S Prajapati, Ujjval B Trivedi and Kamlesh C Patel, 'Kinetic and Thermodynamic Characterization of Glucoamylase from *Colletotrichum* sp. KCP1' (2014) 54 Indian Journal of Microbiology 87.

B. The Patent

The Danish Pharmaceutical Company Novo Nordisk filed a PCT application PCT/DK1998/000520 titled "Thermostable Glucoamylase", claiming
the priority date of 26 Nov 1997. Its corresponding Chinese patent was
granted as CN98813338 (hereafter referred to as the '338 patent).[11] In
2001, the proprietary was transferred to its subsidiary Novozymes, which
is the world's largest provider of industrial enzymes and microorganisms.

The '338 patent disclosed a new type of glucoamylase isolated from a
strain of *Talaromyces emersonii*. This isolated glucoamylase exhibits an
increased thermostability compared to prior art glucoamylases, such as the
Aspergillus niger glucoamylase. It is worth noting that the enzyme in this
invention is not the first glucoamylase that shows thermostability, but a
newly identified one. At 70°C (pH 4.5), the T½ (half-life) was determined
to be over 100 minutes. The specification of this patent disclosed the full
sequence of the thermostable glucoamylase in SEQ ID NO: 7. The relevant claims are as follows:[12]

> *Claim 1: An isolated enzyme with glucoamylase activity, wherein the enzyme*
> **comprises** *the full sequence shown in SEQ ID NO:7.*
> *Claim 6: An isolated enzyme with glucoamylase activity, wherein the enzyme*
> **exhibits a degree of at least 99% identity** *with the amino acid sequence*
> *shown in SEQ ID NO:7, and has a PI[13] below 3.5 determined by isoelectric*
> *focusing.*
> *Claim 10: The isolated enzyme according to claim 6-9 which* **is derived from**
> *a filamentous fungus of the genus Talaromyces, wherein the filamentous fun-*
> *gus is* **Talaromyces emmersonii**.
> *Claim 11: The isolated enzyme according to claim 10, wherein the Ta-*
> *laromyces emmersonii is Talaromyces emmersonii CBS 793.97.*

Claim 1 is to mean that the claimed enzyme has, in its primary structure,
at least the full sequence shown in SEQ ID NO:7. Additional amino acid
residues may exist before or after the reference sequence, which as a consequence extends the scope beyond the disclosed sequence.

---

11 Also granted as a European Patent EP19980958217. See DR Nielsen, RI Nielsen
and J Lehmbeck, 'Thermostable Glucoamylase' <https://encrypted.google.com/pat
ents/EP1032654B1?cl=nl> accessed 10 September 2017.

12 Amended version used in the PRB review, translated by the author. See PRB Decision No. 17956 (31 Dec 2011), <http://app.sipo-reexam.gov.cn/reexam_out/search
doc/decidedetail.jsp?jdh=17956&lx=wx> accessed 10 September 2017.

13 Isoelectric Point (PI): The pH at which the net charge on the protein is zero.

Claim 6 enlarges the scope beyond the reference sequence from a different aspect.[14] It asserts a group of sequences that exhibit "a degree of at least 99% identity" with the reference sequence. "Identity" in this context has a close meaning to similarity or homology. For proteins, it refers to a one-to-one match of the corresponding amino acid residues of the query sequence with those of the reference sequence. A percentage is calculated with a predetermined algorism that defines the penalty scores when there are mismatches or gaps. For example, 100% means two sequences are exactly matching with each other, while 20% shows they are quite different. "Similarity" further concerns residues with similar physicochemical properties, *e.g.* leucine and isoleucine.[15] Hence, for the same set of protein sequences, the degree of similarity can be higher than that of identity. As a keyword of this study, "homology" has its original meaning defined as having shared ancestry in the evolutionary history of life. Strictly speaking, sequence identity/similarity is an observation of two or more given sequences; and homology is the likely conclusion based on a high degree of that. To be scientifically correct, drafters frequently use "identity" or "similarity" instead of "homology". However, unlike "identity" and "similarity", "homology" bears fewer lexicon meanings. The concept is less ambiguous than that of the other two terms when appearing in general contexts. For a clear and concise delivery, "homology" is used in this thesis for a broader meaning embracing both "identity" and "similarity".[16]

Claims 10 and 11 further limit the enzyme mentioned in Claim 6 to be from a particular source. It is derived from the thermophilic fungus *Talaromyces emmersonii,* in particular, from the strain CBS 793.97. Claim 10's limitation narrows down the source of such enzyme to the lowest taxonomic classification, a species. Claim 11 further defines the enzyme from a particular strain stock which is accessible via microbial culture collection centres. A strain is a representative of its corresponding species that

---

14 For the purpose of this thesis, the additional limitation defined by PI is not discussed.

15 Substitution occurred between these amino acid residues are termed conservative substitution, which is generally predicted to have a minimal impact on the tertiary structure of a protein and which thus usually maintains the functionality of this protein. See Simon French and Barry Robson, 'What Is a Conservative Substitution?' (1983) 19 Journal of Molecular Evolution 171.

16 Homology is also used in the field of chemistry, referring to similar functional groups, *e.g.* $-CH_3$ is homologous to $-CH_2CH_3$.

has been collected and preserved or even characterised by the scientific community.

C. The Proceedings on Patent Infringement

Shandong Longda Biology Engineering Co., Ltd. (hereafter, refered to as Longda) and Jiangsu Boli Bioproducts Co., Ltd. (hereafter, referred to as Boli) are major industrial enzyme suppliers in mainland China, both offering thermostable glucoamylase for sale.

In 2011, Novozymes sued Longda and Boli for infringing its patent before the Tianjin Second Intermediate Court (the First Instance Court).[17] In the July of 2011, the two alleged infringer companies filed a Request for Invalidation of the '338 patent to the Patent Reexamination Board (PRB, the Board).[18] In its Decision No. 17956 on 31 Dec 2011, the PRB invalidated some of the claims including Claim 1 and Claim 6, while maintaining the other including Claims 10 and 11.[19]

The Tianjin Second Intermediate Court tried this case based on Claim 10.[20] Novozymes submitted an appraisal conclusion, indicating that the alleged infringing product has the same technical features as Claim 10 in respect of protein sequence and isoelectric point.[21] Novozymes further submitted a search report by the Patent Searching and Consulting Center of the SIPO, indicating that the alleged infringing enzyme cannot originate from organisms other than *T. emersonii*.[22] The alleged infringers failed to prove that the alleged infringing enzyme originated from strains of another species. In its decision, the Tianjin Second Intermediate Court held that Longda and Boli infringed the '338 patent, and awarded Novozymes damages and other fees amounting to CNY 2.2 million (~EUR 270,000 as in

---

17 *Novozymes v Longda*; *Novyzymes v Boli* (n 6).

18 The PRB is the reviewing arm of the State Intellectual Property Office of the Peoples' Republic of China (SIPO). For more procedural requirements for this request, see Yang Zhimin, *New insights on Interlectual Property Law — Detailed Analysis of the Theories and Practice* (知识产权法新解-详析知识产权法的理论与实务) (Sichuan University Press, 2009) 360.

19 PRB Decision No. 17956 (n 12).

20 *Novozymes v Boli*; *Novozymes v Longda* (n 6).

21 Ibid.

22 Ibid.

2012).[23] Longda and Boli appealed. The Tianjin High Court dismissed the appeal and affirmed the decision of the first instance.[24]

Patent invalidity is not an admissible counter-claim in patent infringement proceedings in China's judicial practice. Longda and Boli, thus, had to challenge the patent validity in a separate proceeding. According to Article 11 of *Rules on the Application of Laws in Patent Dispute Proceedings*[25] issued by the Supreme Court, the infringement court may not stay proceedings when the invalidity claim is filed during the defence period. Therefore, in this dispute the infringement was established before the the final decision on the validity of the '338 patent.

D. The Proceedings on Patent Validity

1. The Patent Reexamination Board

The patent litigation in China is bifurcated.[26] The PRB has sole jurisdiction over patent validity issues.[27] In a patent infringement proceeding, the invalidity request must be submitted to the PRB for a review in a parallel proceeding. The '338 patent, which formed the basis of the infringement allegation, was reviewed and held partially invalid by the PRB in its Decision No. 17956.[28] The ground for revocation was Article 26.4 of the Patent Law, which reads as follows:

> *The written claim shall, based on the written description, contain a clear and concise definition of the proposed scope of patent protection.*[29]

---

23 Ibid.
24 *Boli v Novozymes*; *Longda v Novozymes* (n 6).
25 The Supreme People's Court of the People's Republic of China, *Rules on the Application of Laws in the Trial of Patent Dispute Cases* (最高人民法院关于审理专利纠纷案件适用法律问题的若干规定) (19 Jun 2001).
26 Katrin Cremers and others, 'Invalid but Infringed? An Analysis of the Bifurcated Patent Litigation System' (2016) 131 Journal of Economic Behavior and Organization 218. See also Yang Zhimin, *A Study on the Scope of Patent Protection* (Sichuan University Press, 2013) 360 paragraph 1.
27 Patent Law of the People's Republic of China (1984, 2008 Ed.)(the Patent Law) An English version is available at <http://www.wipo.int/wipolex/en/details.jsp?id=5484> accessed 10 September 2017. Article 45.
28 PRB Decision No. 17956 (n 12).
29 The Patent Law (n 27) Article 26.4.

This clause coincides with Article 84 of EPC and Section 112 of U.S. Code Title 35, which is commonly referred to as the *support* requirement in the EU or the *written description* requirement in the USA. In this thesis, *support* will be used in the following text concerning this legal concept.

Claim 1 employs an open-ended transitional phrase "comprise", which encompasses variants that have additional residues before or after the reference sequence.[30] The Board opined that the person skilled in the art would not foresee that adding residues to either end of the reference sequence, by any number and with any type of amino acids, will result in a protein that possesses glucoamylase activity.[31] The Board further gave the following reasons. Firstly, this addition could change the tertiary structure of the protein.[32] Secondly, when this addition results in a much longer sequence than the reference, the reference sequence may be folded inward the protein's tertiary structure, and in such senario the protein loses its original functions.[33] Thirdly, additional residues may interact with those in the protein domains of the reference sequence by forming ionic bonds, hydrogen bonds or disulphide bonds, which as a consequence changes or sabotages the protein domains, or leads to loss-of-function.[34] The Board concluded that open-ended Claim 1 was not supported by the description.[35]

Claim 6 relates to a technical solution defined by the combination of homology and function of a protein or polypeptide. However, only two polypeptides, one with the sequence disclosed in SEQ ID NO:7 and one variant shown in SEQ ID NO:34, were verified to possess glucoamylase activity.[36] The PRB opined that a person skilled in the art could not determine that all variants have the alleged function and can achieve the purpose of this invention.[37] The PRB explained that the basis of a protein's functionality is determined by its tertiary structure, which is subject to change by editing the primary structure, *i.e.* the sequence; substitution made to critical residues would significantly alter the tertiary structure and

---

30 PRB Decision No. 17956 (n 12) 15.
31 Ibid.
32 Ibid.
33 Ibid.
34 Ibid.
35 Ibid 16.
36 Ibid.
37 Ibid.

thus the functionality.[38] Without adequate experimental data in the description, those skilled in the art cannot determine which variants within the claimed homology range would work the invention.[39] Novozymes' submission that the common and general knowledge of conservative substitution would enable those skilled in the art to understand the claim, and that the variant SEQ ID NO:34 which is about 99% homologous to SEQ ID NO:7 demonstrated that the claimed homology range was credible.[40] However, the Board rejected this argument. The Board reasoned that SEQ ID NO:34, as confessed by Novozymes during the oral proceeding, was most probably generated from the infidelity of polymerase chain reaction (PCR). It indicated that SEQ ID NO:34 could have originated from the same source.[41] Nevertheless, SEQ ID NO:34 shares above 99% homology with SEQ ID NO:7, not having reached the bottom line of 99%.[42] Moreover, neither did the written description disclose the conserved domains nor was Claim 6 limited only to conserved substitution.[43] Therefore, Novozymes' argument was not accepted. The Board concluded that Claim 6's technical solution relating to homology was not supported by the description.[44]

Claims 10 and 11 define the origin of the enzyme as *Talaromyces emmersonii,* in particular the strain CBS 793.97. In light of the knowledge that organisms in the same species exhibit high similarity in some fundamental features, the PRB held that an active gene with a specific function would normally have only one sequence in organisms of the same species, and its wild-type sequences with very high homology would have the same function.[45] Given that the glucoamylase activity of the enzyme derived from *Talaromyces emersonii CBS 793.97* had been verified in the description, those skilled in the art would foresee that polypeptides derived from *T. emersonii* and exhibiting at least 99% homology are most

---

38 Ibid.
39 Ibid.
40 Ibid.
41 Ibid. Note that being generated by PCR infidelity does not disqualify SEQ ID NO:34 as a different sequence. This argument seems to have no impact.
42 Ibid.
43 Ibid.
44 Ibid.
45 Ibid 17.

likely to have glucoamylase activity.[46] Therefore, the Board concluded that Claims 10 and 11 were supported by the description.[47]

Claims relating to DNA sequences are not discussed in this thesis, as they are technically connected to protein claims. It is worth noting that nucleic acids and proteins or polypeptides may share some similar arguments, but they do have differences.

2. The Courts of First Instance and Second Instance

Given the infringement decision in the first place, Longda and Boli had to invalidate the patent in its entirety. According to Article 46.2 of the Patent Law[48], they may take legal action against the PRB's decision before a court, more precisely the Beijing First Intermediate Court[49]. On the other hand, although the infringement decision could rely on Claim 10, Novozymes nevertheless wished to recover its patent right related to Claim 6. Consequently, all the three companies filed administrative proceedings against the PRB regarding its Decision No. 17956[50].

The Beijing First Intermediate Court upheld the PRB's decision on the invalidity of Claim 6,[51] and further invalidated Claim 10, Claim 11 and other claims.[52] With regards to Claims 10 and 11, the Beijing First Intermediate Court did not acknowledge the effect of limitation by the species of origin. In the court's opinion, the species of origin limitation did not overcome the defect of allowing random mutagenesis within the defined

---

46 Ibid.

47 Ibid.

48 The Patent Law (n 27), Article 46.2: "A person that is dissatisfied with the patent review board's decision on declaring a patent right invalid or its decision on affirming the patent right may take legal action before a people's court, within three months from the date of receipt of the notification. The people's court shall notify the opposite party in the invalidation procedure to participate in the litigation as a third party."

49 The Beijing Intellectual Property Court took over the first instance from Nov 2014 onwards.

50 PRB Decision No. 17956 (n 12).

51 *Novozymes v PRB,* The Beijing First Intermediate People's Court (2012) Yi Zhong Zhi Xing Chu Zi No. 2596.

52 *Boli v PRB,* The Beijing First Intermediate People's Court (2012) Yi Zhong Zhi Xing Chu Zi No. 2721; *Longda v PRB* The Beijing First Intermediate People's Court (2012) Yi Zhong Zhi Xing Chu Zi No. 2722.

homology range. Thus, the claims still encompassed a huge number of variants, of which the functionality was unpredictable. In conclusion, Claims 10 and 11 lacked support from the written description.[53]

Unsatisfied with either result, Novozymes appealed to the Beijing High Court. The Beijing High Court affirmed all the decisions of the lower court.[54] Regarding the species of origin limitation, the Beijing High Court ruled that "originated from a certain species" did not effectively limit the number of sequences from any organisms within this particular species. Thus such limitation could not cure the defect of a homology claim.[55]

As per the Administrative Procedure Law, this was the in-principle final instance.[56]

3. The Supreme Court

By Article 92.2 of the Administrative Procedure Law, the Supreme Court has the power to hear further appeals and retry cases "where the application of laws and regulations in the original judgment or ruling was truly incorrect".[57] Novozymes thus appealed to the Supreme Court as a last resort on Claim 10, Claim 11 and related claims.

The Supreme Court reasoned in its *Novozymes* decision that: a *species* is a basic unit of biological classification and a taxonomic rank, individuals of which exhibit a high level of similarity in certain fundamental features. The genetic sequence of an enzyme from the same fungal or bacterial species is usually definite, though a very limited number of variants with high homology may exist. Accordingly, the corresponding enzyme is also definite or has only very few variants. The Supreme Court finally held that the double limitations of "at least 99% homology" and species of origin ensured a rather narrow scope of protection, and *a fortiori* Claims 10 and 11 had limitations of enzymatic activity and the isoelectric point

---

53 Ibid.
54 *Novozymes v PRB* (n 51); *PRB v Boli* (n 52); *PRB v Longda* (n 52).
55 *PRB v Boli* (n 5); *PRB v Longda* (n 5).
56 The Administrative Procedure Law of the People's Republic of China (1990, 2015 Ed.) Article 6: "In handling administrative cases, the people's courts shall, as prescribed by law, apply the systems of collegial panel, withdrawal of judicial personnel and public trial and a system whereby the second instance is the final instance."
57 Ibid Article 91.4

followed from Claim 6.[58] Hence, the Supreme Court concluded that Claim 10 and Claim 11 were supported by the description.[59]

Eventually, the Supreme Court upheld the PRB's Decision No. 17956[60] and put an end to the five-year-long dispute on the validity of the '338 patent. In a nutshell, a "homology plus function" claim does not enjoy an easy support from the written description; further experimental data may be demanded; Additional limitation by species of origin will overcome the *support* problem, due to the limited variants and similar functions within a defined species. Now it is known as the function-homology-source rule in the patent law practice in China.

E. Comments – a Good Will, but also a "Chicken Rib"

By recognising the limitation of species of origin, the Supreme Court overcomes the argument that the homology claim encompasses too many unpredictable variants. Given the clear-cut infringement decision delivered in the first place, invalidation of Claims 10 and 11 would have helped the two domestic companies escape from liability if the Supreme Court had not reversed it.

This decision of the Supreme Court may partly reflect China's determination to advance its IP systems and stature. As explicitly expressed by the Premier of the State Council: "The Chinese government sees the fruits of innovation equally; be it by foreign or domestic entities, we provide the same level of protection. The government is dedicated to enhancing the IP protection and strives to build a transparent, fair and just legal and market environment."[61] This ambition has been documented in *Opinion on Accelerating the Building of IP Power under New Conditions*[62], emphasising

---

58 *Novozymes* (n 4) 41.
59 Ibid 42.
60 PRB Decision No. 17956 (n 12).
61 Situ Yuqian, "Premier Li Keqiang Meets WIPO's Director General Gurry" (李克强会见世界知识产权组织总干事高锐) (Beijing, 11 July 2014) <http://www.gov.cn/guowuyuan/2014-07/11/content_2716177.htm> accessed 10 September 2017.
62 State Council of the People's Republic of China, *Opinion on Accelerating the Building of IP Power under New Conditions* (国务院关于新形势下加快知识产权强国建设的若干意见) Guo Fa [2015] 71 <http://www.mof.gov.cn/zhengwuxinxi/zhengcefabu/201512/t20151223_1626379.htm> accessed 10 September 2017.

the importance of IP rights as a means to incentivise innovation and promote economic growth.

This policy concern was also acknowledged by the patentee. "The Supreme Court's decision is of important significance. It shows that China highly values IP protection, which serves as a flag leading the direction of encouraging technological innovation. We believe that the respect towards IP rights will encourage investments in R&D, which is conducive to social development and progress" said Mikkel Viltoft, the general counsel of Novozymes.[63]

This case for the first time clarified the admissible claim styles for patents involving biological sequences, as well as the scope of protection.[64] It provided useful guidance on future drafting and examination practices. For the above reasons, *Novozymes* was enlisted in *TOP10 IP Cases decided by Chinese courts in 2016* and *Typical Cases in Administrative Litigations*.

Be great as it may, the author opines that the significance of this case is limited. As an essential characteristic of biological sequences, the homology issue has not been adequately addressed in *Novozymes*. The concept of homology is almost inevitable in many bio-patents, sometimes it is the sole effective way to describe an invention. Homology not only matters to the support requirement but also serves as a critical factor in other patentability aspects. Novelty, as a substantive requirement of patentability, requires a new biological sequence to be searched against the database. If it is novel and the claimed function is hypothetical, the requirement of industrial application will be assessed with known functions of homologous sequences;[65] or, if it is novel and the claimed function is experimental, the inventive step requirement will be assessed with homologous se-

---

63 Zhu Wenming, 'After the Twists and Turns, Novozymes' Protein Patent Is Finally Maintained'(历经曲折, 诺维信蛋白质专利终被维持) *China Intellectual Property News* (15 March 2017) <http://sipo-reexam.gov.cn/pub/wwzwcsz/alzx/dxalbd/20764.htm> accessed 10 September 2017.

64 Wu Wenying, 'effective limitation by microbial species of orgin in patent claims' (微生物来源限定的权利要求的合理概括) *China Intellectual Property News* (3 May 2017)

65 See T 1452/06, Serine protease/BAYER, EPO Technical Board of Appeal, 10 May 2007.

quences having the same function.[66] High homology increases the possibility of fulfilling industrial application requirement but endangers this molecule regarding the requirement of inventive step, and *vice versa*. Similarly, the inventive step also has something to do with the support requirement. They together delineate the boundaries between prior art and a new invention, as well as between this new invention and a future one. Therefore, the requirements of inventive step and support must coordinate with each other. Without a thorough clarification of homology in the patent law, several problems may arise or continue in future practices.

Firstly, the species of origin is not an effective limitation in the infringement analysis. In the first infringement proceeding, the court admitted the proof that the alleged infringing product was falling within Claim 10 in respect of protein sequence and isoelectric point. Moreover, another piece of evidence indicated that the alleged infringing product could not originate from any other species. It is noted that the court, in order to assure the additional limitation - species of origin, required the alleged infringers to show that the product originated from other sources. Although they failed to prove so, it would be interesting to make a thought experiment for a further discussion: what if the alleged infringers had successfully proved that the product was from another source? This could be a scenario where the two enzymes are identical in any other material features than the species of origin.[67] Would the court have held the case differently? If a counter-proof is meaningful in this scenario, it will render the claim useless in the case of sequences that are highly conserved across species. Alternatively, one can argue that since the enzymes are identical in their material features and the species of origin only makes a difference conceptually, the court would still find infringement.[68] If so, why did the court seek evi-

---

66 See PRB Decision No. 120691 (2 Mar 2017) <http://app.sipo-reexam.gov.cn/reexam_out/searchdoc/decidedetail.jsp?jdh=120691&lx=fs> accessed 10 September 2017. See also T 0111/00, Monokine/FARBER, EPO Technical Board of Appeal, 14 Feb 2002.

67 See Annex I, adapted from Arthur N. Strahler, *Science and Earth History: The Evolution/creation Controversy* (Prometheus Books 1987) *e.g.* Cytochrome c in pig, cow and sheep are identical.

68 Possibly the infringement can be found though the doctrine of equivalents. Note that the doctrine of equivalents is recognised by the Supreme Court. See The Supreme People's Court, O*n Several Issues concerning the Application of Law in the Trial of Patent Infringement Dispute Cases II* (最高人民法院关于审理侵犯专利权纠纷案件应用法律若干问题的解释 (二)), Fa Shi [2016] 1 Article 8.

dence from the alleged infringers in the first place? To make things more complicated, one should bear in mind that an infringing product is usually different from the exact form of the patented invention. It is also probable that the product differs from any known sequences within the defined species. Does it mean that this variant is out of the scope of protection automatically? Or, does it mean that the plaintiff can attribute such variant to that species on certain scientific grounds? If yes, what are those grounds? The only answer is by sequence comparison and homology analysis. In conclusion, the species of origin limitation seems only useful to uphold the validity of a claim, and will not have meanings in the infringement analysis.

Secondly, this case may create an unclaimable gap in biotechnology under the patent law. In *Novozymes*, the '338 patent seems to be limited to a very narrow scope. Although the Supreme Court was only obliged to interpret the law in relation to Claims 10 and 11, it did not make any further comments on homology issues even as *obiter dicta*. It leaves the understanding of homology in the patent law as it was. Unlike the common law jurisdictions, there is no jurisprudence in the Chinese judicial system. Albeit true, it should be noted that this case is an administrative litigation and the PRB was one party in the litigation, meaning that the decision will have a direct influence on the practice of the PRB, and in turn the Patent Office[69]. Therefore, this case, in fact, has a precedence effect on general patent law practices in China. Being important confers this case an exemplary effect on patent drafting and examination, which almost ensures a narrow scope of homology claims in the future. However, the scope of protection is not an isolated concept in patent law; it may interact with the inventive step. An inventive step enables a new invention to escape from the reach of persons skilled in the art. It certainly surpasses the scope of protection of any patents. But what would happen, if the gap between the scope of protection and the inventive step is significantly large? If the scope of protection is narrow, will the bar for of inventive step be lowered accordingly, or will it maintain the status quo? Will there be any problem? In view of these questions, we see that the homology issue is not a problem stirring only the support requirement. It may be intertwined with other legal requirements, which makes the understanding of homology a com-

---

69 State Intellectual Property Office of the Peoples' Republic of China (SIPO).

plex task. From the Supreme Court's decision, no corresponding concerns were reflected.

Thirdly, this case mingles the sufficient disclosure and support requirements. A patent is drafted to enable persons skilled in the art to implement the invention, and to deter the potential infringers who want to circumvent the invention with little meaningful efforts. Consequently, on the one hand, it is at the core of the *quid pro quo* of the patent system to ensure the claimed inventions are workable, and it is also important to draw a fence to repel free-riding attempts, on the other. This ideology indicates that the purposes of the above two are distinct. These two purposes are safeguarded by the sufficient disclosure requirement and the support requirement, respectively.[70] The author's view is that there is an inacceptable merger of the tests on these two requirements during the invalidity proceedings. The support requirement seems to have been tested in the same way as required by the sufficient disclosure requirement. As can be seen from the reasoning of the above proceedings, the large population of variants and poor predictability formed the focus of the debate. From the PRB to the courts, none managed to escape from this topic. Each of them seemed to always have in their mind one simple question – which sequence works? Eventually, by reasoning the limited number of variants within a given species and a high likelihood of similar functionality among them, the Supreme Court was able to confirm the validity of Claims 10 and 11. Considering the prior literal infringement judgement, as long as the Court finds a way to uphold the '338 patent, its policy objective can be achieved. From the perspective of providing general guidance, however, this practice is of limited value when the reasoning was as simple as being written in the decision. The intermingling of the sufficient disclosure and support requirements may continue.

To sum up, *Novozymes* shows a good will of the Chinese judicial system in building a healthy and strong IP environment. However, under scrutiny, it appears to be a "chicken rib"[71] in the real sense – flavourous but fleshless.

---

70 Moreover, another mechanism the doctrine of equivalents also supplements the support requirement for the latter purpose. For the doctrine of equivalents, see note (n 68).

71 See Luo Guanzhong, *Romance of the Three Kingdoms* (XinXii-GD Publishing 2016) Chapter 72; See also the biography of Yang Xiu, available at <http://kongmi ng.net/novel/sgyy/yangxiu.php> accessed 10 September 2017.

III. Homology as an Indication of Confidence

A. Supporting Data for Homology Claims is Not Necessary for the Patent Law

A biological invention usually includes proteins or nucleic acids as its integral components. The building blocks of proteins and nucleic acids are residues of the small composite molecules - amino acids and nucleosides. Therefore, the combinational order of these residues, *i.e.* the sequence, is the precise description of a relevant protein or nucleic acid. Unlike mechanical inventions in which a structural element can be described as "a handle" or "a pad", this kind of language will always be taken as functional rather than structural in a biological invention, given the existence of a more basal description at the sub-molecular level.

The functionality of a biological sequence is subject to its combinational orders. For nucleic acids, these orders are recognised by transfer RNAs to determine the corresponding amino acids, or form hybrids, hairpins and loops to initiate or terminate certain biological processes.[72] For proteins and polypeptides, these orders are the very basis of their activity sites and three-dimensional structures. Any residue can be potentially critical to the function in question, though usually only a few are truly decisive.

Take two examples with relatively simple settings. To investigate the protein exporting mechanism of a bacterium through its Type Four Secretion System, Annette C. Vergunst *et al.* conducted serial mutations to 17 of the last 30 amino acid residues on the C-terminus of a bacterial protein VirF.[73] For their single mutations, four sites were found showing reduced exporting activity by 50%, the exact value of which was also subject to substituting residues. Double mutations based on these four further suppressed the activity to as close as 0%. It can be seen from this example that the function in question (exporting a fusion protein through a secre-

---

72 *e.g.* amiRNA; antisense RNA; guide RNA of the CRISPR-Cas9 system.
73 Annette C Vergunst and others, 'Positive Charge Is an Important Feature of the C-Terminal Transport Signal of the VirB/D4-Translocated Proteins of *Agrobacterium*' (2005) 102 Proceedings of the National Academy of Sciences 832 <http://www.pnas.org/cgi/doi/10.1073/pnas.0406241102> accessed 10 September 2017.

tion apparatus) remains relatively unchanged during most modification efforts. In some cases, a switch from one function to another needs only a tiny modification. For instance, Armin Djamei *et al.* demonstrated a single mutation to mimic constitutively phosphorylated status of the plant protein VIP1.[74] The phosphorylation status is decisive in this protein's subcellular localisation, which in turn affects its subsequent biological events. In a sense, this change could direct to very different technical effects in terms of patent law. A single mutation to mimic phosphorylation constitutively turns on a tunable function to A;[75] if this site is substituted by other residues, the phosphorylation may never occur, thus the function is directed to B permanently. These two examples serve as an appetiser of how sequence-function is correlated. It is this correlation that gives rise to the argument that a homology claim needs back-up by experimental data to show how sequence-function is precisely correlated for the patented invention.

In the author's view, the demand for supporting data is not well grounded. The purpose of the patent law is, as stated in Article 1 of the Patent Law, "to encourage invention-creation and promote the application of invention-creation".[76] From the Paris Convention[77] to the TRIPS Agreement[78], the patent law is always in the commercial context. Meanwhile, an invention is only recognised by the patent law from the technological perspective. An invention was positively defined in the Patent Law as "new technical solutions proposed for a product, a process or the improvement thereof".[79] Again, as stated in Article 7 of the TRIPS Agreement: "The protection and enforcement of intellectual property rights should contribute to the promotion of technological innovation and to the transfer and dissemination of technology". Therefore, as long as an invention carries out its objective technologically and is useful in an industrial sense, it

---

74 A Djamei and others, 'Trojan Horse Strategy in *Agrobacterium* Transformation: Abusing MAPK Defense Signaling' (2007) 318 Science 453 <http://www.science mag.org/cgi/doi/10.1126/science.1148110> accessed 10 September 2017.

75 "Constitutively" means that the said function of that protein becomes constant.

76 The Patent Law (n 27) Article 1, note: "invention-creation" is coined to include inventions, utility models and designs.

77 Paris Convention for the Protection of Industrial Property (1883, as amended on September 28, 1979).

78 Agreement on Trade-Related Aspects of Intellectual Property Rights (1994) (TRIPS Agreement)

79 The Patent Law (n 27) Article 2.

should be acknowledged by the patent law. Accordingly, the *quid pro quo* of the patent law should confer reasonable protection on this invention. Detailed understanding of why and how the invention works like that remains in the scientific realm, to which the patent law could contribute but not as the primary goal.

Scientific advance and technological progress are twins both favoured by a state's policy. They nevertheless bear distinct meanings, and should not be mistaken for each other. Take the famous quote as an analogy: "Humans lit fires for thousands of years before realising that oxygen is necessary to create and maintain a flame."[80] The control of fire has been considered to be a turning point in the cultural aspect of human evolution.[81] It resulted in significant expansion of human activity, and remarkably enhanced the survivability of humanity. None of these great aspects was compromised for not knowing the later-discovered "high-temperature exothermic redox chemical reaction".[82] It is precisely because of the importance of fire that scientists started to investigate the nature of fire. As a tool to promote the application of innovation, the patent law's purpose of encouraging the dissemination of technology will eventually create an eagerness for scientific knowledge underlying certain important inventions. At that stage, more efforts and resources will flood into the scientific investigation; and most probably an answer is revealed by quite a different person from a very distinct field. Imposing such a duty on the patentee or applicant would possibly impede the dissemination of the invention or deter future incentives to make an invention, which in turn harms the development of science. Thus, the patent law should not excuse itself from dealing with tough questions like the homology claims, by diverting the applicants to explore scientific discovery.

The patent in dispute has successfully identified one specific enzyme that is thermostable and has glucoamylase activity. The technical teaching of the invention is complete in terms of the patent law. Although more experiments can be performed to investigate the conserved motifs, domains and even tertiary structures of the enzyme, it is up to the patentee's choice

---

80 *EMI Group North America v. Cypress Semiconductor*, 268 F.3d 1342 (Fed. Cir. 2001).
81 David Price, 'Energy and Human Evolution' (1995) 16 Population and Environment 301.
82 Stephanie R. Dillon, 'The Chemistry of Combustion' <https://www.chem.fsu.edu/chemlab/chm1020c/Lecture 7/01.php> accessed 10 September 2017.

and should not be an obligation. The obligation belongs to the PRB and the courts to think twice about their position on homology claims. The demand for supporting data is nothing but a need to fulfil the courts' scientific curiosity.

B. Supporting Data for Homology Claims is an Overwhelming Burden

In the PRB's and the courts' decisions, it is frequently argued that in light of so many possible variants, those skilled in the art cannot reasonably predict which one works the invention. Admittedly, had *Novozymes* provided enough data in the patent application, they would not have gone through a hard battle over the validity of the claims. From the author's perspective, the provision of substantial data for a homology claim is only a choice of the patentee in theory, but not a doable job in reality.

To understand why the patentee would abandon such a chance to describe the invention more thoroughly, let us take a little test of mathematics. The patented enzyme has 591 residues in its sequence. At least 99% homology means that ≤ 5.91 residues can be substituted, which indicates at most five residues, at any position. The Beijing First Intermediate Court, however, stated that "six residues" is also an acceptable meaning of this homology claim, shown in their reasoning - "up to 5-6 residues can be changed".[83] This opinion was conceded by the Beijing High Court, without doubts. Under this setting, the calculation of combinations are as follows:

$$C(591,1) = \frac{591!}{1!(591-1)!} = 591 \tag{1}$$

$$C(591,2) = \frac{591!}{2!(591-2)!} = 174345 \tag{2}$$

$$C(591,3) = \frac{591!}{3!(591-3)!} = 34229735 \tag{3}$$

$$C(591,4) = \frac{591!}{4!(591-4)!} = 5031771045 \tag{4}$$

$$C(591,5) = \frac{591!}{5!(591-5)!} = 590729920683 \tag{5}$$

---

83 *Boli v PRB* (n 52); *Longda v PRB* (n 52).

$$C(591,6) = \frac{591!}{6!(591-6)!} = 57694622253373 \qquad (6)$$

If the bottom line is set at five substitutes at most, the number of combination is the sum of equations (1) to (5); for six substitutes, the sum of (1) to (6). So, the number of combinations are **5.96×10^{11}** and **5.83×10^{13}**. Moreover, a substitute in any of the five or six sites can be any other amino acid except for the original one.[84] It makes an additional multiplication factor of 19, giving the final numbers as **1.12 × 10^{13}** for five and **1.10 × 10^{15}** for six. This calculation is based on "at least 99% homology" to an enzyme with "591 residues", which is a very narrow range and a normal-sized protein. The figure would be remarkably larger for a wider homology range and a larger protein.

We can see that a mere discretion of including an additional residue (5 → 6) into the interpretation of "at least 99% homology" enlarges the number of combinations by 100 times. Strictly speaking, ≤5.91 is not supposed to include the digit six, either in a legal context or in a technical context. However, the courts seemed not having been well informed before exercising their discretion. This demonstrated an insufficient understanding of the relevant technology when they made those reasonings, which renders their other seemingly sound reasonings questionable.

The SIPO Guidelines only require the patentee to exemplify the derived proteins or polypeptides in accordance wih the claimed homology.[85] But this relaxed standard has little chance to survive through the courts' current argument. In essence, the doubt about a homology claim lies in the large population of variants and the lack of knowledge of which one works. It is not something that can be easily fixed with several examples. It will be sarcastic if the PRB and the courts change their attitudes towards the same claim, just because several examples were provided. This is because in front of the astronomical number of variants, any quantity of examples effectively equals to zero in the proportion. Examples do not make any remedy to the argued problem. Therefore, provision of examples according to the SIPO Guidelines does not logically correspond to the PRB's

---

84 In practice, concerns about properties of the substituted residue will help to reduce the number of candidates.

85 SIPO, *Guidelines for Patent Examination* (2010 Ed.) (the SIPO Guidelines) 355-357. Note, page numbers accord with the English version, available at <http://www.sipo.gov.cn/zhfwpt/zlsqzn/sczn2010eng.pdf> accessed 10 September 2017.

and the courts' argument. Against such argument, exhaustive experimentation is still needed.

There are of course other approaches to facilitate reducing the workload. To identify motifs and domains, alignment of sequences of interest can help to preliminarily predict the conserved regions of the query sequence. However, during the validity proceedings, the sequence alignment data submitted by Novozymes was not well acknowledged. Subsequently, substitutions still need to be carried out to verify the conserved regions. It does not mean that the workload will be automatically lower if possible conserved regions are located. Exactly opposite to this, since the purpose is to support a homology claim, identification of conserved regions may render the patentee in an even awkward situation because when a region is believed to be conserved, the substitutability is thought to be largely limited by those skilled in the art. It means that by identifying one region as conserved, the patentee gives up the possibility that it is in reality not so. Persons skilled in the art will exclude those substitutions[86] made on the alleged "conserved region" from the teaching of the patent. To avoid losing the scope of protection, the patentee would still have to test the substitutions in an exhaustive manner.

The patentee may also resort to structural biology to solve the three-dimensional (3D) structure of the protein, which provides the closest linkage of the sequence-function correlation: sequence - 3D structure - function. That gives the patentee the perfect manner to describe the patented enzyme. However, the correlation of sequence-3D structure becomes another possible question. The courts were reluctant to recognise sequence alignment data as an assumption of conservation during this validity proceeding. There is unlikely to be any difference for a software-based prediction of the 3D structure. Therefore, in an infringement litigation, the patentee need to solve the 3D structure of the alleged infringing goods for the purpose of producing evidence of infringement. Is it a favourable solution? Possibly yes, if the structural biology service is fast and affordable, and if the courts recognise the evidence from computer-facilitated modelling. But the author has doubts over the status quo of its convenience for a litigation, in terms of cost, time, and accountability.

---

86 Here refers to non-conservative substitution, for conservative substitution see French and Robson (n 15).

In a nutshell, the requirement of supporting data to a homology claim is an overwhelmingly high burden to the patentee. Such requirement will greatly retard the disclosure of new inventions. Without fulfilling it, and most probably not being able to discharge it, a patent can only get a very narrow scope of protection based on its homology claim. This consequence is disproportionate to the technical contributions made by the patentee.[87] The relentlessness in emphasising the importance of experimental data in support of homology claims shows a lack of in-depth analysis of the relevant technology and its relation to the patent law. In light of the above, it is necessary for the PRB and the courts to rethink the meaning of homology language and review their positions on homology claims.

C. Rethinking the Role of Homology Language

1. The Homology Language

Broadly speaking, homology claims are not only limited to those with the words of "identity/similarity/homology". Molecule "hybridisation" and "substitution, deletion or addition" bear the same concept. Hybridisation usually relates to nucleic acids. It refers to the thermodynamic phenomena of two nucleic acid strands annealing together by hydrogen-bond formation between bases from each other strand. Hybridisation is a qualitative description, it can be predicted even without experimentation. Because of the thermodynamic nature and quantitative understanding of chemistry, the strength of hybrid (usually expressed as the Tm; melting temperature) is easily calculated under a given ionic strength environment. Yet, this description is more qualitative than quantitative, as it usually gives an answer of "yes" or "no" under low/moderate/high stringent conditions. Moreover, different numbers of hydrogen bonds formed between A/T and C/G[88] means that the impact of various substitutions may also differ ac-

---

87 Although a protein-related invention is disclosed with its sequence, the actual teaching is that one "category" of proteins perform the mentioned function. The principle of the function by such "category" of proteins remains the same, as understood by the persons skilled in the art. See William R Pearson, 'An Introduction to Sequence Similarity ("homology") Searching' [2013] Current Protocols in Bioinformatics.

88 Two hydrogen bonds formed between A/T, A=T; three between C/G, C≡G.
   See J Berg, J Tymoczko and L Stryer, *Biochemistry* (2007) p112.

cordingly. Simply put, changing A or T to others results in a minor alteration in the Tm compared with changing C or G. It makes a prejudice over C and G substitutions in the sequence, which may not have grounds in terms of its biological meanings.[89] "Substitution, deletion or addition" is an operational description, but can be combined with quantitative elements like "substitution, deletion or addition with one or several residues". This type of language has an advantage of delivering the homology concept of the relevant claim to a layperson, because of its description from an operational aspect. Comparing the different types of homology claims, "substitution, deletion or addition" and homology are better choices. They do not distinguish which residue is substituted, and they define the precise relative proximity to the reference sequence in a quantitative manner.

2. The Technical Meaning of Homology

A molecule is the smallest physical entity for a chemical compound that has the chemical properties of that compound. Insofar according to such understanding, a molecule might not be accepted by only partial description, since the partial description of a molecule is not conclusive of the final properties of this molecule. This could be the underlying principle that the PRB did not accept "comprising" or "contain" as a method to describe a protein, though the PRB argued this point based on numerous variants. The requirement of full sequence description brings about one problem that the patentee or applicant will be responsible for any of those "unimportant parts" in the claimed molecules. As argued by the PRB, the rest parts of a molecule may sabotage the claimed function in several ways.[90] At this point, the PRB started to question the enzymatic activity in the claims – the functional limitation. This idea followed in when the analysis continued to the homology claims. Thus, the sequences that are homologous to the disclosed one suffer from the same problem of unpredictability, with regards to their functionality.[91]

---

89 When the function of the nucleic acid in question is to form hybrids or hairpins, this prejudice is justified, as the function *per se* is hybridisation; but when the nucleic acid molecule carries further information, like mRNA or DNA, this prejudice has no justification.
90 PRB Decision No. 17956 (n 12) 15. See Section II.D.1.
91 PRB Decision No. 17956 (n 12) 16. See Section II.D.1. See also Pearson (n 87).

However, the challenge on non-functional sequences is not appropriate in the patent law practice. There are two possible ways to understand the mentioned function in the claims[92] of the '338 patent: 1) the function is part of the title of the subject matter; or 2) the function is used as a functional limitation. Whichever understanding is chosen for the interpretation, it must serve to limit the claims. According to O*n Several Issues concerning the Application of Law in the Trial of Patent Infringement Dispute Cases II*, both preamble portion and characterising portion have the limitation effect. [93] If this rule is followed, those homologous sequences that do not perform the mentioned function will fall outside the claims. For this reason, it is not necessary to make emphasis on the side of non-functional sequences when assessing the validity. Instead, the assessment should be focused on whether those functional ones are claimable.

When a particular sequence with any of its functions is disclosed, the persons skilled in the art do not merely believe that only this specific sequence performs the mentioned function. Rather, those skilled persons will understand that such function can be carried out by sequences similar to the disclosed one.[94]The only ambiguous thing in this understanding is how similar they should be. At this point, the homology range indicated in the claims serves as the basis of their confidence. It does not define absolutely whether any single homologous sequence within this homology range should work, but provides a cut-off value based on which a judgement is less likely to be wrong. The understanding of homology language as a matter of confidence can be demonstrated in the following examples.

In a research article, the authors state that:

> *In view of the strong conservation of the Skp1 proteins, we predict that Skp1 proteins of other plant species such as N. glauca will interact with VirF similarly to the way in which A. thaliana Skp1 homologs do.*[95]

---

92 See Section II.B: "an isolated enzyme with glucoamylase activity".

93 See note (n 68) Fa Shi [2016] 1: Article 5.

94 See Vineet Sangar and others, 'Quantitative Sequence-Function Relationships in Proteins Based on Gene Ontology' (2007) 8 BMC Bioinformatics 294 <http://bmc bioinformatics.biomedcentral.com/articles/10.1186/1471-2105-8-294> accessed 10 September 2017.

95 Barbara Schrammeijer and others, 'Interaction of the Virulence Protein VirF of *Agrobacterium tumefaciens* with Plant Homologs of the Yeast Skp1 Protein' (2001) 11 Current Biology 258.

The "strong conservation" refers to a high homology of the said protein "Skp1" across "plant species" other than "*A. thaliana*". We can see from this statement that 1) the authors had no hesitation in nominating similar proteins from other species as Skp1; and 2) the author made clear and unambiguous statement that untested Skp1 proteins from other species can be predicted to function in the same way. Similar views not only exist in scientific publications, but also appear in patent law cases. In the European Patent Office (EPO) case **T0111/00**[96], the claimed cytokine[97] is of human origin. The closest prior art disclosed the sequence of a cytokine of mouse origin. The claim was held to be obvious as 1) there is an incentive to find the claimed cytokine based on the prior art cytokines; and 2) using the mouse cytokine cDNA as a probe to isolate human cytokine is straightforward.[98] A decision for lacking an inventive step demonstrates that the Technical Board of Appeal (TBA) believed that finding the claimed cytokine required no inventive efforts in light of the known cytokines. This belief of the TBA is based on homology between the claimed and the known cytokines. And this belief concurs with the author's opinion of homology as a confidence indicator. These two examples show that the concept of homology indicates a confidence level, which in turn projects the prospect of conducting further research on a particular sequence having such homology.

D. Species of Origin is Not an Effective Limitation

It is worth noting that according to the above two examples, the confidence level acquired from the knowledge of homology did not stop at the boundary of species classification, nor did it stop at the border of genus.[99]

---

96 T 0111/00 (n 66).

97 A category of small proteins that function in cell signalling. See Charles A. Dinarello, 'Historical Review of Cytokines' (2007) 37 European Journal of Immunology S34.

98 The straightforwardness is understood as such: in light of high homology, DNA hybridisation can be reasonably predicted. Therefore, the cDNA of the known protein, or part of it, can be used to probe unknown but suspected homologous DNA from other sources.

99 "Plant" is a kingdom-level description, six levels higher in taxonomic ranking than species; mouse and human converge in the class level – mammalia, four levels higher in taxonomic ranking than species.

Using species of origin as a limitation is thus not reflecting the essence of the technical facts. The Court supported such limitation for the reason that it effectively limited the number of naturally existing variants.[100] This can only be understood as an *ad hoc* solution in front of the patent in dispute, and should not mean that species of origin is a good and the only acceptable limitation.

Species by definition is the largest group of organisms in which two individuals are capable of producing fertile offsprings.[101] They do share some fundamental features, features that are functional to ensure reproduction. No underlying logic can be found to generally ensure proteins from the same species function in the same way, especially an industrially applicable one whose function is determined by human objectives. What if the claimed function is to cause blood precipitation when mixed with anti-A serum? The isoagglutinogen found in human body can work very differently: some people's isoagglutinogen can cause blood precipitation in anti-A serum, some cannot. This is the well-known case of blood cross-matching test for ABO blood group system.[102] Therefore, The Court was wrong in arguing, in its *Novozymes* decision, that within a particular species, there are only several sequences for a protein, and they share the same functions.[103] The only benefit in a species of origin limitation is the limited number of naturally existing variants. Therefore, the Court's acceptance of the species of origin is not well-grounded.

Moreover, when species of origin is used as a limitation, highly conserved sequences may encounter problems. When two sequences from different species are identical,[104] technical aspects of the two sequences will most probably be the same. As a consequence, this type of limitation may embarrass the courts in deciding whether it constitutes an infringement. If no infringement is found, it will technically make such patent non-en-

---

100 Though the large number of variants should not be a proper perspective in the validity discussion, the Court seemed not able to escape from this argument. This point will be further discussed in the following section.

101 See *Species* at Wikipedia <https://en.wikipedia.org/wiki/Species>.

102 See Fumiichiro Yamamoto, 'Review:ABO Blood Group system—ABH Oligosaccharide Antigens, Anti-A and Anti-B,A and B Glycosyltransferases, and ABO Genes' (2004) 20 Immunohematology 3.

103 *Novozymes* (n 4) 41. See Section II.D.3.

104 See Annex I (n 67). See also University of Missouri-Columbia, 'Identical DNA Codes Discovered in Different Plant Species' <https://www.sciencedaily.com/releases/2012/04/120409164426.htm> accessed 11 September 2017.

forceable; if infringement is found, probably through the doctrine of equivalents, [105] the limitation will prove itself only useful for patent validity not for infringement analysis.

In an infringement proceeding, attribution of an alleged infringing product to a certain species could be tricky. From the scientific perspective, molecular identification of a species usually relies on some specific elements in the organism, exemplified by 16S ribosomal RNA, due to the slow rates of evolution of this region of the gene.[106] It is not scientifically sound to identify a particular species based on an arbitrary sequence in a patent dispute. Even if this has to be done, the only ground to determine the origin of a given sequence is to make comparisons with the reference sequence of a particular species, and by assessing the confidence based on the homology value. From the above discussion, it becomes clear that homology is an inevitable consideration.

E. Concluding Remarks

In view of the discussions in this section, the support requirement for homology claims is neither necessary in terms of patent law nor doable in terms of technological and legal practice. A proper understanding should be given to the homology claims, which is not primarily meant to encompass a population of variants but to indicate a level of confidence on the sequence-function correlation. The use of species of origin as an alternative limitation appears to be a good way to discharge the support requirement, but is in fact only effective to stabilise the disputed patent given the improper arguments submitted from the lower courts. So far, this decision has no further value as a guide for future patent practice. Homology claims, thus, should be acknowledged by the patent law as an important method to describe a biological invention, and the support requirement should be reconsidered with a holistic analysis.

In the following sections, the author will analyse homology claims in the larger context of patentability requirements. The self-consistency of

---

105 For doctrine of equivalents, see note (n 68).

106 CR Woese and GE Fox, 'Phylogenetic Structure of the Prokaryotic Domains: The Primary Kingdoms' (1977) 74 Proceedings of the National Academy of Sciences, USA 5088 <http://www.pnas.org/cgi/doi/10.1073/pnas.74.11.5088> accessed 11 September 2017.

these requirements will be assessed against homology claims, and the predicament in applying Article 26.4 of the Patent Law[107] will be reviewed.

---

107   The Patent Law (n 27).

IV. Novozymes may Create an Unclaimable Gap

A. Inventive Step and Support are One-Dimensionally Aligned by Homology

The Patent Law defines the inventive step as involving two separate requirements: prominent substantive features and notable progress.[108] An invention has prominent substantive features when it is not obvious to the person skilled in the art.[109] This non-obviousness is assessed against the technical motivation of the person skilled in the art to apply the different features on the closest prior art.[110] The other requirement, the notable progress, is to mean the advantageous technical effects.[111] The SIPO Guidelines enumerates four criteria to fulfil this requirement, in which No. 2 recognises that a different inventive concept to achieve substantially the same technical effect in the prior art qualifies notable progress.[112] An invention like in the '338 patent has its inventive concept as providing a new type of enzyme that performs prior art functions. In other words, this kind of invention finds a different way to produce a technical effect in the prior art. The requirement of notable progress is thus fulfilled in such scenario. In the following discussion, the inventive step will be identical with non-obviousness.

The inventive step confers upon a patented invention a distance beyond the reach of persons skilled in the arts, and the support requirement tunes the claimable scope of protection. For sequence-related inventions targeting the same technical effect, an independent patentable invention in assessing its inventive step must supersede the claimed scope of protection of the prior art patent and its extent of obviousness. The relationship between a later independent patent (B) and the prior art patent (A) will be as expressed below:

---

108 The Patent law (n 27) Article 22.3. Note that the Patent Law employs "creativity" for this concept.
109 The SIPO Guidelines (n 85) 195.
110 Ibid 196, Section 3.2.1.1 (3).
111 Ibid 200, Section 3.2.2.
112 Ibid.

Inventive Step B (p2) > Claimed Scope A (p1) + Obviousness (p2)   (7)

In this inequation, the left side represents the later invention, whereas the right side stands for prior art patent. The numerals accompanying each term refer to different persons skilled in the art. To determine the claimed scope of the prior art patent (A), the relevant date for those skilled is the filing date (p1). As for the inventive step of a later patent (B), those skilled in the art have the relevant date of filing that patent (p2). The obviousness of the prior art patent (A) is thus assessed on the same date (p2).

Thereafter, if the Claimed Scope A is to the best extent a patentee can claim, it equals Claimable Scope A. It means that there is no such fault of the patentee that she gives up any scope if she can satisfy the support requirement. In the situation of homology claims, as discussed in Section III.B, the patentee faces tremendous variants and has no effective way to counter an argument like which one works. There is nothing that the patentee is intending to give up, as coined by the concept of the doctrine of dedication.[113] The inequation transforms to:

Inventive Step B (p2) > Claimable Scope A (p1) + Obviousness (p2)   (8)

Now we assume that the scientific knowledge of the sequence in the prior art patent (A) has not evolved, and the mutagenesis techniques have no evolutionary progress against large scale experimentation. Under this assumption, the common and general knowledge of the persons skilled in the art are stable for p1 and p2. The only difference between the two will be that p2 and p1 may have different dates to determine prior arts. This assumption can be tested against the history of prosecution and invalidation proceedings of Patent B. If no other prior art relevant to the sequence or technical argumentation is raised, this case falls into the hypothetical scenario. In such a scenario, Obviousness A effectively approaches zero, as long as Patent B and Patent A deals with the same technical effects. Because, person p2, without any new knowledge and better skills, cannot add anything more when the earlier patentee has claimed to the best extent - Claimable Scope A.

---

113 The Supreme People's Court, O*n Several Issues concerning the Application of Law in the Trial of Patent Infringement Dispute Cases* (最高人民法院关于审理侵犯专利权纠纷案件应用法律若干问题的解释), Fa Shi [2009] 21 Article 5.

Now the inequation becomes:

Inventive Step B (p2) > Claimable Scope A (p1)   (9)

p2 has its relevant date on the filing date of Patent B, and p1 has the date of filing Patent A. But what exactly is the difference between p2 and p1? The assumption leading to (9) already dictates that no new knowledge and better skills are relevant. And certainly she (p2) knows the prior art patent (A). What does p1 know? Given the statutory language "based on the written description"[114], it is apparent that she should have known the teachings of Patent A, otherwise she has nothing to base upon when drafting claims. The only difference is that p2 reads the claims that p1 has written, thus p2=p1. Taken that they are persons with the same knowledge and the same skills, they should be able to reach an acceptable consensus to delineate the claimable and free-to-operate spaces. To this point, the inequation has been simplified to the following:

Inventive Step B > Claimable Scope A   (10)

From the above reasoning, we know that Claimable Scope A is governed by the support requirement. If the patentee did the best she could to safeguard her interest, the Claimable Scope is then solely a matter of discretion of the support requirement. This discretion, in the context of this thesis, is the attitude on homology. Given that Patent B performs the same technical effect as Patent A does, the only aspect to establish an inventive step is how the way of doing so is different. For a biological sequence, the sole criterion is how far is deemed to be a safe distance to separate two sequences as different inventions. Again, this is a decision on homology. So far as homology is concerned, the requirements of inventive step and support are now aligned one-dimensionally.

B. Disparity in Views on Homology Creates an Unclaimable Gap

In the preceding discussion, some preconditions are set forth to reveal the interrelation of inventive step and support requirements. They are: 1) the inventive concept only covers the identification of a particular sequence of

---

114   The Patent Law (n 27) Article 26.4.

a new kind, and does not extend to a new technical effect; 2) no further knowledge relating to the prior art sequence enters the public domain; and 3) there is no revolutionary technological progress in dealing with a large scale experimentation. Condition 1 and 2 can be found by examining the prior art documents produced in a proceeding when assessing the inventive step; Condition 3 is by default given, in view that the question "which one works" and the scale of experimentation makes a very high burden. In this section, another PRB case is analysed to examine the PRB's position on homology for the requirement of inventive step.

The patent application CN 201080053990 by Novozymes disclosed polypeptides having xylanase activity and their coding polynucleotides, isolated from *Penicillium pinophilum*. The priority date was 29 Sep 2009. One piece of prior art, dated in 2005, disclosed a xylanase from *Penicillium funiculosum*, with both the amino acid sequence and the coding sequence.[115] Through sequence alignment, one of the claimed amino acid sequence SEQ ID No:2 shares 96.56% homology with the prior art sequence.[116] The examination division rejected relevant claims for lack of inventive step. The examination division reasoned that in light of high homology, the finding of another xylanase within the same genus was obvious.[117] In the appeal, the PRB maintained the decision and added more detailed technical reasonings: persons skilled in the art could clone the highly homologous sequence, using primers designed from the prior art sequence. The test of enzymatic activity was known, and the prior art sequence provided enough motivation to find a functional enzyme in a species within the same genus. In conclusion, the application was rejected for lack of inventive step.[118]

This decision suffers from some flaws in its reasoning. As the author found, the genome sequence of *P. pinophilum* was only made available be-

---

115 Caroline SM Furniss, Gary Williamson and Paul A Kroon, 'The Substrate Specificity and Susceptibility to Wheat Inhibitor Proteins of *Penicillium funiculosum* Xylanases from a Commercial Enzyme Preparation' (2005) 85 Journal of the Science of Food and Agriculture 574.

116 The nucleic acid sequence has a homology of 92.79% compared to the prior art. But the homology value of nucleic sequence and amino acid sequence cannot be judged using the same standard. Because, codon degeneracy allows certain level of changes to the nucleic sequence without troubling the person skilled in the art.

117 The actual finding work usually bases on the homologous nucleic acid sequence.

118 PRB Decision No. 120691 (n 66).

tween 2016 and 2017, [119] the "high homology" between the claimed sequence and the sequence in the cited prior art was unlikely to be known by the inventor before hand. It is only after the claimed sequence had been successfully identified by the inventor that 96.56% homology became relevant. Before the invention was created, the existence of a high homology itself remained a hypothesis. Therefore, the argument based on a later-identified high homology is not an appropriate reasoning,[120] though the same decision may be reached in other ways. Despite all this, the decision sends out a clear signal that homology is such a prominent factor that the examiners and the Board largely rely upon it. Furthermore, in evaluating the inventive step, the functionality of a sequence did not attract much attention; the PRB explained that as long as the assumption of functionality was there, to test the functionality with known methods would be a routine task.

As an anecdote, it is interesting to know that during the review, the homology range of the claimed sequence was amended from 99% to 100%. It means that only the disclosed sequence is sought for protection, and no real homology claim is involved. A species of origin limitation is already included in the original version. The complete abandonment of homology claim took place during the proceedings of Novozymes' glucoamylase patent, after the Beijing High Court's ruling and before the decision of the Supreme Court. It is apparent that the invalidation decision from the first two instances affected the applicant's confidence on such kind of claims. Since this xylanase patent application is also from Novozymes, the same applicant centred around in this thesis, it makes the xylanase patent a

---

119 Cheng-Xi Li and others, 'Genome Sequencing and Analysis of *Talaromyces pinophilus* Provide Insights into Biotechnological Applications' (2017) 7 Scientific Reports 490 <http://www.nature.com/articles/s41598-017-00567-0> accessed 10 September 2017. Note that the species name was changed: *Talaromyces pinophilus = Penicillium pinophilum* <http://www.mycobank.org/BioloMICS.aspx?TableKey=14682616000000067&Rec=480573&Fields=All> accessed 10 September 2017.

120 The SIPO Guidelines (n 85) 209, Section 6.2: "when evaluating the inventive step of an invention, the examiner is apt to underestimate the inventive step of the invention since he has already known the contents of the invention, and hence a mistake of *ex post facto* analysis is likely to be made. Therefore, the examiner shall always bear in mind that, in order to reduce and avoid the influence of subjectivity, the evaluation shall be presumed to be made by a person skilled in the art on the basis of comparison between the invention and the prior art before the filing date thereof".

quintessence to demonstrate a plight as such: a patent applicantion cannot support its homology claims, even in a very narrow range of ≥99%, unless species of origin is further limited; but a lower homology like 96.56% has no problem penetrating the boundaries of species classification, and rendering an invention obvious.

Adding to the discussion on species of origin limitation, we now see that species of origin can help to reduce the doubt of support about homology, but cannot prevent the influence of confidence by homology. This phenomenon makes the species of origin only a passive choice of an applicant, but should not be a justifiable and universal method to further limit a homology claim.

Although the xylanase patent received an inventive step rejection based on a 96.56% homology, it is still not clear how low a homology should be to escape such rejection. In examples provided by the Japanese Patent Office (JPO),[121] Case 6 shows a scenario where the claimed sequence shares 80% homology with a prior art sequence, and it is held lack of an inventive step, unless the difficulty to obtain the claimed sequence can be otherwise provided. The earlier mentioned EPO case T 0111/00 showed that 78% homologous to the prior art sequence made the claimed sequence obvious. But the obviousness was partially based upon secondary considerations.[122] Considering other patent offices' practice, it is thus reasonable to believe that in China the actual threshold of homology to establish an inventive step can be much lower than the exemplified value. In view of the JPO's example and the EPO's case law, the threshold in China has no reason to be above 90%.

When incorporating the homology values into Inequation (10), we see that to establish inventive step, the homology is supposed to be much lower than 96.56% - possibly the requirement will not be any easier than 90%; and 99% cannot get supported without species of origin, which effectively puts the current value of support in fact at 100%. As a consequence, a large gap appears between the homology thresholds of support and inventive step requirements. A problem then arises - what is the nature of the unclaimable gap?

---

121 Japan Patent Office, *Examples of examinations on the inventions related to genes* <http://www.jpo.go.jp/cgi/linke.cgi?url=/tetuzuki_e/t_tokkyo_e/dnas.htm?url=/te tuzuki_e/t_tokkyo_e/dnas.htm> accessed 11 September 2017.

122 T 0111/00 (n 66).

C. The Unclaimable Gap May Constitute a Discrimination

The unclaimable gap may find its justification based on the fact that the person skilled in the art usually searches for new inventions among sequences in their wild-type form; but to support a homology claim, the patentee has to consider all other possible mutations.

In the review of the xylanase patent, the PRB correctly pointed out that the person skilled in the art can use primers to probe the possible homologous sequences in species within a certain taxonomic classification. Although the author argued that the existence of such homologues and their homology are all in hindsight, it does not change the fact that the skilled persons have a relatively small pool to conduct their searching. Therefore, the belief on possible homology can effectively lead the person to a claimed sequence. Meanwhile, in the support requirement, the PRB and the courts wished to apply a similar argument of the pool size. They reasoned that in light of the large population of variants and the lack of sequence-function knowledge, a skilled person could not predict which variant works. Should this argument be justifiable, the unclaimable gap might find its grounds. But under scrutiny, this is unlikely to be the case.

The searching in the wild-type has its root in reality. It is survival of the life in their natural environments that gives rise to the different sequences and their functions. The knowledge of any particular sequence having any useful function owes largely to the naturally existing. Even in the era of synthetic biology, when the skilled have the power to edit these sequences to work in an entirely different way, this correlation is still unbreakable. Unlike the situation where a person can apply the laws of nature in an arbitrary technical embodiment, in the sequence-related invention man gets the idea from a sequence and embodies an invention as such, or a new idea into another sequence. This makes a man-made sequence in any matter a mimic or an alteration of the natural ones. Therefore, for any existing function already achieved by prior art sequences, the searching for a new type, albeit a homologous one, is always conducted among the natural sequences. This is a confidence beyond the sequence-function correlation; it even extends to the existence of such sequences. But considering that there is a limited number of natural ones to test, this confidence justifies an attempt.[123]

The situation in supporting a claim is quite different. The purpose of claims is to prevent misappropriation. The primary goal of the claimed homology is to defend against arbitrary modifications.[124] In such a case, a homology claim has to face a tremendous amount of variants due to the combinations exemplified in Section III.B. If the same argument for inventive step is applied here, there is no chance to discharge the requirement. However, the seemingly numerous variants are the perspective from the applicant. While the purpose of the claim is to defend against other parties, the standing point should not be of the applicant but a person who wishes to achieve the same technical effect starting from the disclosed sequence, especially a wild-type one. From this person's perspective, the aim is to find one working variant, not to test every single variant in the claimed range. In light of the confidence based on high homology, this person expects to experience only a few trials before she reaches one that works perfectly to achieve the same technical effect.[125] In this scenario, the experimentation burden for this person cannot be deemed high.

The knowledge of a sequence is predominantly a matter of top-down discovery, not a bottom-up design of an inventor. An inventor can only contribute to combining and altering certain functions, but the building blocks remain as a gift of nature. Therefore, there is little chance for the person skilled in the art to reach a similar sequence without knowing the one in prior art. In other words, a variant is not an independent creation,

---

123 Following the Court's in argument *Novozymes* (n 4) that there is usually one or several sequences among individuals of the same species, the inventor only need to try representative strains of one species, and the amount to examine is largely dependent on the availability of sample species.

124 T 2101/09, Human Delta3 Notch/MILLENNIUM, EPO Technical Board of Appeal, 26 Feb 2013. "It is common practice in the field of biotechnology that claims [...] are not required to be limited to a very specific sequence but may also embrace molecules having a certain degree of homology and/or identity to this specific sequence. [...] This practice allows patentees/applicants **to protect their inventions against arbitrary modifications of the specific sequences**".

125 Conversations with two biotechnology researchers provided the expected number of trials as follows: a technician in AppTec (a global research outsourcing provider in Wuxi, China) projected less than 10 possible trials before reaching a working variant, based on "99% homologous to a 591 AA sequence"; a postdoctoral researcher in Singapore Agency for Science, Technology and Research (A*STAR) predicted 20 possible trials at most.

but a derivative. Hence, a change made to the claimed sequence without targeting any other technical effects can never contribute to technology. This effort could never be deemed inventive, and should not be encouraged. However, if the change is done to generate other functions, it will automatically fall outside the claimed scope of protection, in light of the further functional limitation in a homology claim.

A certain level of homology has conferred upon the skilled persons a confidence both to conduct searching and to make working variants. As discussed in the preceding paragraphs, the huge amount of variants is only a fact from an omniscient perspective, not the perspective of the skilled persons. From the skilled person's perspective, the claimed homology range is not a "laundry list"[126] which seeks an extension of protection to those not tested, but a boundary which prevents non-inventive and non-meaningful efforts in modifying the claimed sequence. Thus, it is not appropriate to impose an unreasonable burden upon an applicant claiming a homology range. The unclaimable gap in Chinese patent law practice seems ungrounded.

Admittedly, the support requirement need not always match the standard of inventive step. There is always a possibility that further knowledge and techniques infiltrate into the public domain, and push the inventive step further. But it is not the case discussed in this thesis. Analysis of the relevant knowledge and techniques has already ruled out the contribution from other sources. This thesis enjoys a privilege to align only the support and inventive step requirements in a single dimension so that a significant mismatch is conspicuously exposed.

Protection in exchange for disclosure forms a fundamental principle of the patent law.[127] Under this principle, the system of patent law works in such way as to grant a term of monopoly on the economic aspects of an invention, making the technological contributions eventually fall into the public domain. The unclaimable gap, however, paves another way to directly put an inventor's contributions into the public domain without any compensation measures. The unclaimable gap, in its nature, is a direct de-

---

126 See, *e.g. Fujikawa v. Wattanasin*, 93 F.3d 1559, 1571, 39 USPQ2d 1895, 1905 (Fed. Cir. 1996) "a 'laundry list' disclosure of every possible moiety does not necessarily constitute a written description of every species in a genus because it would not 'reasonably lead' those skilled in the art to any particular species".

127 T 1452/06, Serine protease/BAYER, EPO Technical Board of Appeal, 10 May 2007 para 23. See also, *Pfaff v. Wells Electronics, Inc.* 525 U.S. 55 (1998) 63.

privation of the technical contributions made by an inventor. There are many ways that the public domain can benefit from additional technological progress without delaying, by publication of books and articles, the obviousness in light of prior art combinations, the doctrine of dedication and the abandonment of patent rights. But expropriation of protection from a patentable invention should never be one of them. Such conduct will undermine the purpose of the patent law. By creating a gap, this practice prevents an inventor either from claiming a reasonable scope of protection, or from establishing an inventive step with the matching standard. As a result, technical contributions within such a gap directly fall into the public domain.

This phenomenon in the patent practice is nothing but a *de facto* discrimination to biotechnology, which is not tolerated by the TRIPS Agreement. As required by Article 27.1 of this Agreement:

> ...*[P]atents shall be available and patent rights enjoyable without discrimination as to ... the field of technology...*[128]

Unlike the common form of discrimination, which prevents patentability of inventions from certain technological fields,[129] the discrimination in homology claims does not refuse protection but sets up unfriendly double-standards that expropriates some technical contributions of applications and patents. In view of the ambition to build up a strong IP environment, this inappropriate practice should be corrected.

D. Downregulating Inventive Step is Not a Feasible Option

The unclaimable gap, without a plausible cause from the growth of prior art and common and general knowledge, is a result of decoupling persons skilled in the art. For this reason, either side needs to be examined against their proper capability. To restrict the unclaimable gap, two options can be made. One is to lower the bar for inventive step, and the other is to relax the support requirement.

---

128 TRIPS Agreement (n 78).

129 Stefania Fusco, 'TRIPS Non-Discrimination Principle: Are Alice and Bilski Really the End of NPEs?' <https://papers.ssrn.com/sol3/papers.cfm?abstract_id=2653463> accessed 10 September 2017.

But, to lower the bar for inventive step may not be an feasible option. Firstly, being a substantive requirement of patentability, the requirement of inventive step has been explicitly included in Article 27.1 of the TRIPS Agreement.[130] On the other hand, the requirement of support lacks grounds in international treaties. A treaty-level requirement is supposed to perform a role in international patent law harmonisation. It influences proper functions of international filing cooperations. The requirement of support is dealing with the drafting of claims which thus enjoys more flexibility with regards to amendment, compared with written descriptions. In light of the exemplified views from JPO and EPO, it is unlikely that the practice in China will change. Secondly, lowering the bar for inventive step may lead to the tragedy of anticommons.[131] This option inevitably avails more patents surrounding the first known sequence-function correlation. In its appearance, it looks as if an implementer has multiple choices. However, with the expansion of relevant knowledge and techniques, these closely situated patents may grow in their equivalent powers.[132] Possibly, significant merger of scope will occur among multiple patents. At that moment, one particular functional sequence may face multiple right owners. The exploitation in turn becomes extremely difficult. This is exactly an unfavourable situation typified by the tragedy of anticommons. Lastly, even if the significant merger might not occur, a lower inventive step would still be unfavourable, as it finally affirms the narrow scope of protection, disincentivising innovation as no one is likely to receive enough economic reward to recoup their costs or to support further research and development.[133]

130　TRIPS Agreement (n 78) Article 27.1: "patents shall be available for any inventions, whether products or processes, in all fields of technology, provided that they are new, **involve an inventive step** and are capable of industrial application".

131　MA Heller and Rebecca S Eisenberg, 'Can Patents Deter Innovation? The Anticommons in Biomedical Research' (1998) 280 Science 698 <http://www.science mag.org/cgi/doi/10.1126/science.280.5364.698>.

132　The relevant date for the doctrine of equivalents is the date of infringing activity. See The Beijing High Court, *Guidelines for Patent Infringement Determination* (2013) Article 44.

133　See Kenneth G Chahine, 'Enabling DNA and Protein Composition Claims: Why Claiming Biological Equivalents Encourages Innovation' (1997) 25 AIPLA QJ 333.

The author thus seeks to address a plausible solution to the unclaimable gap on the support's side.

V. Novozymes Mingles Sufficient Disclosure and Support

A. Sufficient Disclosure and Support Have Different "Prior Art"

The current interpretation on homology claims not only rooted in an insufficient understanding of the technology, but also came from the failure to distinguish the support requirement from sufficient disclosure. In the patent law, sufficient disclosure and support are two separate requirements. The former is reflected in Article 26.3 of the Patent Law:

> *The written description shall contain a clear and comprehensive description of the invention or utility model so that a technician in the field of the relevant technology can carry it out; when necessary, pictures shall be attached to it. The abstract shall contain a brief introduction to the main technical points of the invention or utility model.*[134]

The latter is stated in Article 26.4 of the Patent Law:

> *The written claim shall, based on the written description, contain a clear and concise definition of the proposed scope of patent protection.*[135]

These two requirements constitute both the grounds for refusal and revocation, in accordance with Article 53.2 and Article 65.2 *Rules for Implementation of the Patent Law*[136]. The two requirements are closely connected. When the breadth of the claim exceeds the technical contributions of an inventor, and some ways within the claimed scope owes nothing to the patent or application to achieve the desired result, both grounds can be invoked for an invalidity challenge. This scenario is usually mentioned as

---

134 The Patent Law (n 27). This article coincides with Article 83 EPC, usually referred as the requirement of sufficient disclosure.

135 The Patent Law (n 27). This article coincides with Article 84 EPC, usually referred as the requirement of support.

136 Rules for Implementation of the Patent Law of the People's Republic of China (2001, 2010 Ed.). An English version is available at <http://www.wipo.int/wipolex/zh/text.jsp?file_id=182267> accessed 12 September 2017. Note, the support requirement is not a ground for revocation under Article 138 EPC.

Biogen insufficiency,[137] and can also be exemplified in the PRB Decision No. 23542.[138]

Nevertheless, they are different. The fundamental difference can be read from the wordings of each provision. Article 26.3 describes the standard for "written description" being sufficiently clear to enable others; while, Article 26.4 states that the "claims" are drafted based on the written description. Apparently, before drafting or reading the claims, the teaching of an invention is already laid down in the written description. The skilled addressees when informed of the asserted protection will firstly bear the teaching of the written description in mind. Therefore, the desired results are not merely a matter starting from the prior art of that patent or application, but must be considered in combination with the teachings already disclosed in the written description. As a consequence, whether not directly disclosed matters are protectable may not have the same assessment with sufficient disclosure.

B. Novozymes Tests Support Using the Standard of Sufficient Disclosure

To be sufficient, the written description teaches a person, from the beginning, how to work the invention. This requirement corresponds to the fact that being inventive this patent should advance beyond the reach of persons skilled in the art on the filing date. However, when the skilled persons have been enabled to work this invention in the disclosed way(s), any other ways to work the invention should be able to refer to the disclosed one(s).

In view of such difference, it is important to understand the relationship between the first sequence that qualifies the sufficient disclosure and those other sequences homologous to the first one. To this point, Robert Hodges argued that "the key event is the cloning of the first gene in a family of corresponding genes. Once a researcher accomplishes this very difficult task, the researcher can typically obtain other members of the gene family

---

137 See *Biogen v Medeva* [1997] RPC 1 HL. Note, according to Article 138 EPC, support is not a ground for revocation. Therefore, a challenge for this reason should employ the ground of insufficiency.

138 PRB Decision No. 23542 (23 July 2014) < http://app.sipo-reexam.gov.cn/reexam _out/searchdoc/decidedetail.jsp?jdh=23542&lx=wx> accessed 11 September 2017.

with much less effort."[139] This idea has been adequately reflected in the assessment of inventive step. But in the support requirement, it seems to have been ignored. Following Hodges' logic, when the first sequence and its function is disclosed, the search for other functional homologous sequences "is conducted on the basis of what is known, that is, the function, rather than on the basis of what is unknown - the precise structure" said Burk and Lemley.[140]

The first sequence provides the very initial but fundamental idea that "a particular sequence can do a certain kind of job". Subsequently, looking for the other variants having the same function, no matter whether naturally existing or arbitrarily modified ones, will be significantly easier. This fact reveals that, the major technical contribution originates from the identification of a particular sequence-function correlation. Although detailed information may be lacking as on what basis or to what extent the sequence is tolerant to alteration, it only amounts to a minor concern in comparison to the contribution. Particularly, the vast majority of variants encompassed in a given homology range are arbitrarily modified; and considering the top-down nature of the knowledge in biotechnology, they are impossible to be reached without referring to the disclosed sequence. As long as the distance in homology is reasonably close, even without detailed knowledge, persons skilled in the art will have no problem in predicting the similar function of a variant.

Skilled persons in sufficient disclosure and support may need to answer the same question when the claim contains parallel methods or products which are not fully described and which do not share the same fundamental technical principle. But it is simply not the case for sequence-related inventions. For a sequence-related invention, the first sequence-function

---

139 Robert A Hodges, 'Black Box Biotech Inventions: When a "Mere Wish or Plan" Should Be Considered an Adequate Description of the Invention' (2001) 17 Georgia State University Law Review <http://readingroom.law.gsu.edu/gsulr> accessed 10 September 2017. See also John M Lucas, 'The Doctrine of Simultaneous Conception and Reduction to Practice in Biotechnology: A Double Standard for the Double Helix' (1998) 26 AIPLA Quarterly Journal <http://heinonline .org/HOL/Page?handle=hein.journals/aiplaqj26&id=389&div=16&collection=jo urnals> accessed 10 September 2017.
140 Dan L Burk and Mark A Lemley, 'Is Patent Law Technology-Specific?' (2002) 17 Berkeley Technology Law Journal 1155 <https://login.e.bibl.liu.se/login?url=h ttps://search.ebscohost.com/login.aspx?direct=true&db=aph&AN=9133378&site =eds-live&scope=site> accessed 10 September 2017.

correlation forms the singularity. Whatever population the rest of the sequences may be, they only amount to a tiny fraction compared to the technical contribution.

These two persons thus face very different questions. The person in sufficient disclosure is about to read the written description. He may want to know how to obtain the said molecule from available sources, what is the identity of this molecule – the sequence, what is the specific utility, and if not available in the prior art yet – verification methods and technical parameters of its function. The person in support will examine that, given an arbitrarily drafted homology range, whether within this distance of homology a variant is still believed to perform a similar function. The first person needs comprehensive and credible information, whereas the second person does not have to answer the question as right or wrong, but an answer as more likely than not.

Therefore, the question to the person in the support requirement should be like a "likelihood of success" in assessing the inventive step. The role of a high homology can be exemplified into two scenarios: 1) if Seq A is known, Seq B is highly homologous to A, there will be a good estimate that Seq B has the same function with Seq A; and 2) if Seq A is known, and someone wishes to modify Seq A within a reasonable homology range, she will experience only limited trials to reach a certain Seq B which maintains Seq A's function. These two scenarios are rooted in the same level of confidence if the value of homology is set. The only difference is whether Seq B is given or to be found. The accuracy of estimation in scenario 1) thus negatively corresponds to the difficulty of finding a Seq B the scenario 2). When 1) is clearly held obvious, it indicates that 2) is not an undue burden.

Having discussed all the above, let us review the reasoning given by the PRB and endorsed by the courts: "without adequate experimental data in the written description, those skilled in the art cannot determine which variants within the claimed homology range, other than the disclosed, would work the invention."[141] Apparently, the PRB wants a concrete answer, which assumes that the skilled person is still so naïve that she needs further teaching to be enabled. But the fact is that the singularity is already reached in the written description, and she is anyway enabled in the first place. The person in support now should, in turn, assess within what dis-

---

141 PRB Decision No. 17956 (n 12) 16.

tance the variants cannot escape the gravity of the technical contribution of the first enablement.

C. An Example Test Given by the EWHC

Having found that the PRB asked an inappropriate question to the person in the support requirement, a correct question needs to be exemplified for future direction.

In *GlaxoSmithKline v Wyeth*[142], GlaxoSmithKline (GSK) wished to clear the way for its vaccine Bexsero, and sought to challenge the validity of Wyeth's UK part of European patent EP2343308 [143] on multiple grounds. And Wyeth counterclaimed for infringement of the patent. The Claim 1 is as follow:

> *A composition containing at least one protein comprising an amino acid sequence **having sequence identity greater than 95% to the amino acid sequence** of any one of SEQ ID NOs: 212, 214 and 216, wherein the composition additionally comprises at least one PorA protein.*[144]

As one of those grounds, GSK challenged the threshold figure of 95% homology.[145] GSK argued that this homology threshold did not arise from any of the data in the specification (written description) of the patent, and the figure was arbitrary. Henry Carr J. disagreed GSK's argument, and sided with the technical expert Prof Ala'Aldeen's opinion that "the skilled person would understand from the data [that specified] protein was a useful antigen that elicits antibodies which are cross-bactericidal, and would expect that effect to be related to the degree of amino acid homology."[146] The expert further viewed that "a claim to utility based on 95% homology would be entirely credible and well above the level of homology which would cause the skilled person to question it".[147]

---

142 *GlaxoSmithKline UK Ltd v. Wyeth Holdings LLC* [2016] EWHC 1045 (Ch).

143 GW Zlotnick and others, 'Novel Immunogenic Compositions for the Prevention and Treatment of Meningococcal Disease' <https://www.google.com/patents/EP2 343308B1?cl=en> accessed 10 September 2017.

144 *GlaxoSmithKline v Wyeth* (n 138) [68].

145 According to Article 138 of the European Patent Convention, support is not a ground for revocation, instead GSK challenged on the ground of insufficiency, which in its essence is a challenge of support.

146 *GlaxoSmithKline v Wyeth* (n 138) [104].

147 Ibid.

The reasoning held by Henry Carr J. only dealt with the credibility under the given homology range, and did not touch upon the absolute
question: which one works? Moreover, this credibility was built upon the
known factor that the disclosed sequence was confirmed to be functional
in the first place. This approach corresponds with the author's opinion that
homology claims stand for a level of confidence, not an abrupt inclusion
of the huge amount of variants. The huge amount of variants can only be
interpreted from its native context, the perspective from the skilled persons. The skilled person appreciates no technological advance or plurality
over the number of variants, but the remarkable significance of the first sequence-function correlation. Though needs may arise to find working variants, they never seek to make it exhaustive. Thus, the support requirement
ought to be assessed individually rather than on the whole, *i.e.* can the person skilled in the art, inspired by the disclosed sequence, reach at one
working variant without undue burden? – If yes, the claimed range is supported. By posing the appropriate question to the skilled person in the support requirement, the unclaimable gap will be effectively constrained.

D. On Non-Working Variants – How to Avoid a "Negative Gap"?

Having addressed the working variants, it is still important to analyse the
non-working variants within an asserted homology range. Non-working in
this context does not necessarily mean that a variant has no functionality,
but only indicates that it does not perform the function as mentioned in the
claims.

The author's discussion on the appropriate questions is mainly to address the implicated unclaimable gap. However, there could be a reverse
scenario where the accepted homology range by the support requirement
exceeds a plausible inventive step, *e.g.* a minor mutation resulted in a new
and irrelevant technical effect, for which an inventive step may subsist.
Should this happen, the two requirements may face a crossover in their re-
spective homology values. This in turn appears to create a negative gap.
When the homology claim relates to an absolute product protection, *i.e.*
the product *per se*, a clash in homology values may translate into a prejudice over the later inventive mutation, as the later invention already es-
capes from the technical contribution documented in the first patent.

This "negative gap" exists because of a breach of the preconditions that
the author sets forth in Section IV.A. One-dimensional alignment of sup-

port and inventive step only occurs under the condition of relating to the same technical effect. However, the "negative gap" takes place when the support and the inventive step are regarding different technical effects. Thus, it is important to limit the corresponding protein or polypeptide of the disclosed sequence to a particular technical effect before homology should be employed.

The SIPO Guidelines explicitly addressed that a protein claim could be drafted in such a way:

> *A protein of (a) or (b) as follows:*
> *(a) a protein whose amino acid sequence is represented by Met-Tyr-...-Cys-Leu,*
> *(b) a protein derived from the protein of (a) by substitution, deletion or addition of one or several amino acids in the amino acid sequence in (a) and having the activity of enzyme A.* [148]
> ...

Paragraph (b) includes the homology concept "substitution, deletion or addition" and a value "one or several amino acids". Meanwhile, this claim also includes "having the activity of enzyme A" as a functional limitation. This example suggests that homology only works in combination with a given function. Thus, a homology claim in its entirety includes "homology plus function". The homology language is not meant to work alone. In Section III.C, it is already discussed that homology, in essence, represents a level of confidence, but more importantly a confidence on what? This confidence is about the achievability of a certain goal — in this case, a certain function that the homologous sequences perform.[149] Otherwise, a simple homology description does not generate any technical meaning. Hence, in the alignment of the support and inventive step requirements, the coordinate axis of homology is conditioned by the same function of different sequences. Only under the condition of having the same function, homology values can be coordinated and analysed on along the same dimension. A homologous sequence asserting another technical effect may well find its way in filing an independent patent, without being threatened

---

148 The SIPO Guidelines (n 85) 357.
149 See Sangar and others (n 94). See also UK Biotech Guidelines (n 2) 49, "Claims should be limited by reference to the activity of the reference sequence where there is doubt about the identity of a homologue in relation to the reference sequence".

by the earlier claimed scope of protection. For this reason, the "negative gap" does not exist from the beginning.

In *Novozymes*[150], the enzymatic activity in the claimed homology range, without a further limitation by species of origin, was assigned as a burden for the patentee. This decision, in its effect, conceded that the functional limitation of enzymatic activity has little value in asserting the scope of protection. From the discussion of the preceding paragraphs, it is clear that homology claims must include both elements: homology and functionality. Thus, the current practice broke the homology claim apart, and attenuated the validity of the homology language used in the claims. This practice is neither supported by the SIPO Guidelines,[151] nor by the technical understanding of homology in biotechnology. Therefore, to assess the requirement of support, functional limitation within a homology claim should never be put aside.

---

150 *Novozymes* (n 4).
151 The SIPO Guidelines (n 85) 357

VI. Conclusion

Homology is a crucial concept in sequence-related biotechnological inventions. The homology language in a claim primarily seeks to defend against misappropriation by arbitrary modifications. However, this type of claim faces an insurmountable hurdle before the requirement of support, which currently requires an overwhelmingly high burden of experimental data.

The support requirement concerning homology is not standing alone in the patent law. Via the person skilled in the art, it can be aligned with the requirement of inventive step on the same coordinate axis of homology, when dealing with the same technical effect. Moreover, it should be distinguished from the requirement of sufficient disclosure due to the different knowledge those skilled persons have in these two requirements.

Novozymes implicates an unclaimable gap in biotechnology under the patent law, the formation of which comes from a very narrow allowance in the claimable scope of protection in contrast to a comparably large distance in establishing an inventive step. This gap significantly decoupled the skilled persons in the requirements of support and inventive step, with no relevant prior art or common and general knowledge to blame. Thus this gap may constitute a *de facto* discrimination towards biotechnology.

To restrict the unclaimable gap along homology, either reducing the bar for inventive step or relaxing the requirement for support can be opted for. But the former one is unfavourable and possibly leads to more problems. The latter one finds its grounds in that *Novozymes* implies a misapplication of sufficient disclosure standard onto the support test. The author opines that skilled persons in these two requirements have different knowledge relating to the sequence. The skilled person in the requirement of sufficient disclosure learns *ab initio* the first sequence-function correlation, which forms the most significant part of the patent's technical contributions. However, the skilled person in the support requirement has known the first sequence-function correlation, thus does not need to address the sufficiency *de novo*.

The skilled person in the requirement of support is thus to examine whether it is credible and without undue burden to reach a working variant within the claimed homology range. By pinpointing the claimed homology range as an indication of confidence, the author demonstrates that the skilled person in the requirement of support can evaluate whether obtaining working variants is doubtful or needs undue burden. In line with the proposed approach to assessing the support requirement for homology, functional limitations must be duly acknowledged, not only to facilitate asserting the scope of protection but also to abalienate the rights on embraced sequences to the inventions bearing different inventive concepts.

The patent law safeguards the economic interests of inventors and promotes innovation and social development.[152] Thus, the protection it confers to a patented invention should commensurate to the inventor's technical contribution. What a sequence-related invention contributes is the provision of a general sequence-function correlation. This technical contribution goes far beyond the identification of a particular sequence having a useful function. It also extends to the homologous sequences whose functions become expectable in light of the disclosed sequence-function correlation. Although it is not easy to quantify such contribution into a numerical homology value, the persons skilled in the art do have the ability to examine whether a given boundary is credible.

Lastly, the provision of protection should bear its genuine intention – to protect. When homology effects subsist, a patented invention exposes itself to numerous ways of misappropriation. No matter how forceful the patent is, homology could always be its *Achilles' heel*. If the patent law aims at building a strong shield to protect patentable inventions. Isn't homology the very place where the *Aegis* shall be?

---

152 See Article 1 of the Patent Law (n 27).

Annex I: Sequences of Cytochrome c from 17 Different Species

Amino Acid # ----> 1 ... 10 ... 20 ... 30 ... 40 ... 50 ... 89

- Human
- Rhesus monkey
- Horse
- Donkey
- Common zebra
- Pig, cow, sheep.
- Dog.
- Gray whale
- Rabbit.
- Kangaroo.
- Chicken, Turkey.
- Penguin.
- Pekin duck.
- Snapping turtle
- Rattlesnake
- Bullfrog.
- Tuna.
- Silkworm moth.
- Wheat.
- Baker's yeast
- Neurospora

[CONTINUED FROM ABOVE]

Amino Acid # ----> 60 ... 70 ... 80 ... 90 ... 100 ... 110 ... 112

- Human
- Rhesus monkey
- Horse
- Donkey
- Common zebra
- Pig, cow, sheep.
- Dog.
- Gray whale
- Rabbit.
- Kangaroo.
- Chicken, Turkey.
- Penguin.
- Pekin duck.
- Rattlesnake
- Snapping turtle
- Bullfrog.
- Tuna.
- Silkworm moth.
- Wheat.
- Baker's yeast
- Neurospora

Annex I: Sequences of Cytochrome c from 17 Different Species

Adapted from:
Strahler AN, Science and Earth History: The Evolution/creation Contro-versy, (Prometheus Books 1987)

List of Works Cited

Books and Journal Articles

Berg J, Tymoczko J and Stryer L, *Biochemistry* (2007)

Burk DL and Lemley MA, 'Is Patent Law Technology-Specific?' (2002) 17 Berkeley Technology Law Journal 1155

Chahine KG, 'Enabling DNA and Protein Composition Claims: Why Claiming Biological Equivalents Encourages Innovation' (1997) 25 AIPLA QJ 333

Cremers K and others, 'Invalid but Infringed? An Analysis of the Bifurcated Patent Litigation System' (2016) 131 Journal of Economic Behavior and Organization 218

Dinarello C a, 'Historical Review of Cytokines' (2007) 37 European Journal of Immunology S34

Djamei A and others, 'Trojan Horse Strategy in *Agrobacterium* Transformation: Abusing MAPK Defense Signaling' (2007) 318 Science 453

French S and Robson B, 'What Is a Conservative Substitution?' (1983) 19 Journal of Molecular Evolution 171

Furniss CSM, Williamson G and Kroon PA, 'The Substrate Specificity and Susceptibility to Wheat Inhibitor Proteins of *Penicillium funiculosum* Xylanases from a Commercial Enzyme Preparation' (2005) 85 Journal of the Science of Food and Agriculture 574

Fusco S, 'TRIPS Non-Discrimination Principle: Are *Alice* and *Bilski* Really the End of NPEs?' <https://papers.ssrn.com/sol3/papers.cfm?abstract_id=2653463>

Luo G, *Romance of the Three Kingdoms* (XinXii-GD Publishing 2016)

Heller MA and Eisenberg RS, 'Can Patents Deter Innovation? The Anticommons in Biomedical Research' (1998) 280 Science 698

Hodges RA, 'Black Box Biotech Inventions: When a "Mere Wish or Plan" Should Be Considered an Adequate Description of the Invention' (2001) 17 Georgia State University Law Review

Kumar P and Satyanarayana T, 'Microbial Glucoamylases: Characteristics and Applications' (2009) 29 Critical Reviews in Biotechnology 225

Li C-X and others, 'Genome Sequencing and Analysis of *Talaromyces pinophilus* Provide Insights into Biotechnological Applications' (2017) 7 Scientific Reports 490

Lucas JM, 'The Doctrine of Simultaneous Conception and Reduction to Practice in Biotechnology: A Double Standard for the Double Helix' (1998) 26 AIPLA Quarterly Journal

Marín-Navarro J and Polaina J, 'Glucoamylases: Structural and Biotechnological Aspects' (2011) 89 Applied Microbiology and Biotechnology 1267

Nielsen BR, Nielsen RI and Lehmbeck J, 'Thermostable Glucoamylase' <https://encrypted.google.com/patents/EP1032654B1?cl=nl>

Prajapati VS, Trivedi UB and Patel KC, 'Kinetic and Thermodynamic Characterization of Glucoamylase from *Colletotrichum* sp. KCP1' (2014) 54 Indian Journal of Microbiology 87

Pearson WR, 'An Introduction to Sequence Similarity ("homology") Searching' [2013] Current Protocols in Bioinformatics

Price D, 'Energy and Human Evolution' (1995) 16 Population and Environment 301

Schrammeijer B and others, 'Interaction of the Virulence Protein VirF of *Agrobacterium tumefaciens* with Plant Homologs of the Yeast Skp1 Protein' (2001) 11 Current Biology 258

Strahler AN, *Science and Earth History: The Evolution/creation Controversy* (Prometheus Books 1987)

Vergunst AC and others, 'Positive Charge Is an Important Feature of the C-Terminal Transport Signal of the VirB/D4-Translocated Proteins of *Agrobacterium*' (2005) 102 Proceedings of the National Academy of Sciences 832

Woese CR and Fox GE, 'Phylogenetic Structure of the Prokaryotic Domains: The Primary Kingdoms' (1977) 74 Proceedings of the National Academy of Sciences, USA 5088

Wu W, 'effective limitation by microbial species of origin in patent claims' *China Intellectual Property News* (3 May 2017)

Yamamoto F, 'Review: ABO Blood Group system—ABH Oligosaccharide Antigens, Anti-A and Anti-B,A and B Glycosyltransferases, and ABO Genes' (2004) 20 Immunohematology 3

Yang Z, *New insights on Intellectual Property Law – Detailed Analysis of the Theories and Practice* (Sichuan University Press, 2009)

—— *A Study on the Scope of Patent Protection* (Sichuan University Press, 2013)

Zhu W, 'After the Twists and Turns, Novozymes' Protein Patent Is Finally Maintained' *China Intellectual Property News* (15 March 2017)

International Treaties and Legislations

Agreement on Trade-Related Aspects of Intellectual Property Rights (1994)

Paris Convention for the Protection of Industrial Property (1883, as amended on September 28, 1979)

Patent Law of the People's Republic of China (1984, 2008 Ed.)

Rules for Implementation of the Patent Law of the People's Republic of China (2001, 2010 Ed.)

The European Patent Convention (16th Ed., June 2016)

The Administrative Procedure Law of the People's Republic of China (1990, 2015 Ed.)

Internet Sources

Situ Y, "Premier Li Keqiang Meets WIPO's Director General Gurry" (Beijing, 11 July 2014) <http://www.gov.cn/guowuyuan/2014-07/11/content_2716177.htm>

Stephanie R. Dillon, 'The Chemistry of Combustion' <https://www.chem.fsu.edu/che mlab/chm1020c/Lecture 7/01.php>

University of Missouri-Columbia, 'Identical DNA Codes Discovered in Different Plant Species' <https://www.sciencedaily.com/releases/2012/04/120409164426.htm>

Official Documents

Intellectual Property Office of the UK, *Examination Guidelines for Patent Applications relating to Biotechnological Inventions in the Intellectual Property Office* (6 May 2010, last updated: 21 October 2016)

Japan Patent Office, *Examples of examinations on the inventions related to genes* <http://www.jpo.go.jp/cgi/linke.cgi?url=/tetuzuki_e/t_tokkyo_e/dnas.htm?url=/tetu zuki_e/t_tokkyo_e/dnas.htm>

State Council of the People's Republic of China, *Opinion on Accelerating the Building of IP Power under New Conditions*, Guo Fa [2015] 71 <http://www.mof.gov.cn/zhe ngwuxinxi/zhengcefabu/201512/t20151223_1626379.htm>

State Intellectual Property Office of the People's Republic of China, *Guidelines for Patent Examination* (2010 Ed.)

The Beijing High People's Court, *Guidelines for Patent Infringement Determination* (2013)

The Supreme People's Court of the People's Republic of China, *Rules on the Application of Laws in the Trial of Patent Dispute Cases* (19 Jun 2001)

The Supreme People's Court, O*n Several Issues concerning the Application of Law in the Trial of Patent Infringement Dispute Cases*, Fa Shi [2009] 21

The Supreme People's Court, O*n Several Issues concerning the Application of Law in the Trial of Patent Infringement Dispute Cases II*, Fa Shi [2016] 1

Cases

EPO

T 0111/00, Monokine/FARBER, EPO Technical Board of Appeal, 14 Feb 2002

T 1452/06, Serine protease/BAYER, EPO Technical Board of Appeal, 10 May 2007

T 2101/09, Human Delta3 Notch/MILLENNIUM, EPO Technical Board of Appeal, 26 Feb 2013

China

PRB Decision No. 17956 (31 Dec 2011) , <http://app.sipo-reexam.gov.cn/reexam_out/ searchdoc/decidedetail.jsp?jdh=17956&lx=wx>

PRB Decision No. 23542 (23 July 2014) < http://app.sipo-reexam.gov.cn/reexam_out/s earchdoc/decidedetail.jsp?jdh=23542&lx=wx>

PRB Decision No. 120691 (2 Mar 2017) <http://app.sipo-reexam.gov.cn/reexam_out/s earchdoc/decidedetail.jsp?jdh=120691&lx=fs>

Novyzymes v PRB, The Beijing First Intermediate People's Court (2012) Yi Zhong Zhi Xing Chu Zi No. 2596

Boli v PRB, The Beijing First Intermediate People's Court (2012) Yi Zhong Zhi Xing Chu Zi No. 2721

Longda v PRB, The Beijing First Intermediate People's Court (2012) Yi Zhong Zhi Xing Chu Zi No. 2722

Novozymes v PRB, The Beijing High People's Court (2014) Gao Xing (Zhi) Zhong Zi No. 3522.

PRB v Longda, The Beijing High People's Court (2014) Gao Xing (Zhi) Zhong Zi No. 3523.

PRB v Boli, The Beijing High People's Court (2014) Gao Xing (Zhi) Zhong Zi No. 3524.

PRB & Novozymes v. Boli, the Supreme People's Court (2016) Zui Gao Fa Xing Zai No.85

PRB & Novozymes v. Longda, the Supreme People's Court (2016) Zui Gao Fa Xing Zai No.86

UK

Biogen v Medeva [1997] RPC 1 HL.

GlaxoSmithKline UK Ltd v. Wyeth Holdings LLC [2016] EWHC 1045 (Ch)

US

Pfaff v. Wells Electronics, Inc. 525 U.S. 55 (1998)

EMI Group North America v. Cypress Semiconductor, 268 F.3d 1342 (Fed. Cir. 2001)

Patents

Nielsen BR, Nielsen RI and Lehmbeck J, 'Thermostable Glucoamylase' <https://encrypted.google.com/patents/EP1032654B1?cl=nl>

Kariyone A, Hashizume Y and Hayashi R, 'Enzyme Electrode for Measuring Malto-Oligosaccharide and Measuring Apparatus Using the Same' <http://www.google.com/patents/EP0335167A1?cl=en>

Zlotnick GW and others, 'Novel Immunogenic Compositions for the Prevention and Treatment of Meningococcal Disease' <https://www.google.com/patents/EP2343308B1?cl=en>